Activated Sludge Technologies
for Treating Industrial Wastewaters

5/15/2015

Matt

To my good friend and colleague at Geosyntec.

Best in future!

[illegible]

HOW TO ORDER THIS BOOK

BY PHONE: 877-500-4337 or 717-290-1660, 9AM–5PM Eastern Time

BY FAX: 717-509-6100

BY MAIL: Order Department
DEStech Publications, Inc.
439 North Duke Street
Lancaster, PA 17602, U.S.A.

BY CREDIT CARD: American Express, VISA, MasterCard, Discover

BY WWW SITE: http://www.destechpub.com

Activated Sludge Technologies for Treating Industrial Wastewaters

Design and Troubleshooting

W. WESLEY ECKENFELDER, JR., D.Sc., P.E.
JOSEPH G. CLEARY, P.E., BCEE

Activated Sludge Technologies for Treating Industrial Wastewaters

DEStech Publications, Inc.
439 North Duke Street
Lancaster, Pennsylvania 17602 U.S.A.

Printed in the United States of America
10 9 8 7 6 5 4 3 2 1

Main entry under title:
Activated Sludge Technologies for Treating Industrial Wastewaters: Design and Troubleshooting

A DEStech Publications book
Bibliography: p.
Includes index p. 217

ISBN No. 978-1-60595-019-8

Table of Contents

Special Dedication

My co-author, Dr. William Wesley Eckenfelder, died on Sunday, March 28, 2010, just as the ideas for this book were being formed. Wes was a pioneer and living legend in the profession and also a great friend and colleague. I had the distinct privilege to work with him on many projects and workshops over the last ten years. A great tribute to Wes was the memorial service held in Nashville on April 1, 2010. Anyone who was there—and many of his former students were—will never forget it or forget Wes. He was the godfather of our industrial wastewater profession for many years.

I have great memories of Wes. In the early 1970s I graduated from Manhattan College's Environmental Engineering Program. Wes graduated from Manhattan in the mid-1940s and taught there for several years. A legend at Manhattan College, he left there in the 1960s for the University of Texas and then moved to Vanderbilt in 1969. I used Wes's textbooks in school, and learned about him from a couple of my professors. Don O'Connor, the professor who really got me interested in Environmental Engineering, was a colleague of Wes's and co-author of his first book in 1960. I missed by about 5 years having Wes as a teacher. At Hydroscience, Inc. I missed Wes, who had left a few years prior to my arrival. Don and Wes started Hydroscience in 1961. Over the years I met Wes a few times at the Manhattan Alumni Plumbers dinners but never really got to know the man behind the reputation. I never imagined I would get to work with him and become friends with him many years later.

It wasn't until the year 2000 when I was recruited to Eckenfelder/Brown and Caldwell that I had the opportunity to get to know Wes. The main reason I took this job over others was to be able to work with Wes. I was always amazed when I met someone who had attended one of his workshops years ago and still spoke about it as a memorable experience.

Wes and I conducted numerous workshops over the last ten years at several key WEF industrial waste conferences and other venues and universities. The first workshop we did was held at the El Conquistador Hotel in Fajardo, Puerto Rico and attended by pharmaceutical industry clients who filled the room.

Wes asked me to substitute for him one year at the Manhattan College's Summer Institute, which I had heard many times at workshops. That was in 2002 and I continued doing the Summer Institute for several years. I spoke with him each year to review how to improve the content of the course. One of the key memories was at the 50th Anniversary of the Summer Institute. Wes was one of the four speakers that Friday afternoon at a major gathering of alumni who were later that evening to attend the Plumbers. Wes, Bob Thomann, Charles O'Melia and Dom DiToro, the four keynote speakers all gave presentations. It was a great event in which Wes led off with the 50 years of history of industrial wastewater treatment technology developments. Wes was the only engineer in the business who could do that talk because he was one of the co-founders and even inventors of industrial wastewater treatment. Even though he could not see well, he went through all the slides and received a standing ovation. It may have been the best presentation I have seen Wes deliver. He had the unique ability and talent to take a complex subject and present it so that it was easily understood as well as very entertaining and enjoyable.

After working together for several years and spending time with him on numerous trips I was honored to have Wes as a friend. He was a great colleague and mentor to me on projects, proposals and workshop presentations. I am amazed at how he continued to have the passion for helping clients solve their wastewater issues as well as the passion to teach engineers and students at the workshops, even into his eighth and ninth decades of life. He did all of this with hardly any eyesight left and a need to sit during lectures.

A final recollection for me was in July 2007. Wes received the inaugural Lifetime Achievement Award from the Water Environment Federation at the Industrial Waste Conference in Providence, Rhode Island. No one was more deserving of that award, which was named in his honor. Sometimes I envy the students who were lucky to have him in college and the long-time friends who knew him much longer, but I feel so very fortunate to have had the opportunity late in my career to be close to him as a colleague and friend.

His lifetime of dedication inspired this dedication and the book that follows.

JOSEPH CLEARY

Preface

EXCELLENT books are available on activated sludge biological treatment, which are used by students, plant operators and engineers. Most of these, however, focus on municipal, in contrast to industrial, wastewater. As the activated sludge process turns 100 years old, the authors of this book, after speaking with practitioners, operators and engineers, felt that a new text with a focus on industrial wastewater treatment was needed to cover the following developments and advances:

- the continued evolution of the activated sludge process and its numerous designs, configurations and technology developments;
- more attention to design of industrial water reuse systems following the activated sludge process to achieve industry sustainability goals for water and energy;
- changes over the years from BOD, TSS and nutrient removal to removal of specific organics, toxicity, worldwide concerns about microconstituents, and more stringent effluent permit limits;
- advances in process modeling tools that can be used in combination with treatability testing tools for plant design, optimization and troubleshooting;
- concerns over industrial wastewater discharge impacts to POTWs, such as nitrification inhibition, the impact of frac water from the development of shale gas supplies and the fate of microconstituents through POTWs.

After a century of use, activated sludge is going strong and at the same time changing. Activated sludge remains the backbone or workhorse process in most biological treatment plants, even with decades of technology advancements and more stringent effluent discharge requirements. In the 1950s and 1960s, industrial wastewater treatment primarily concentrated on removal of

BOD and TSS. In the late 1980s and early 1990s, EPA categorical effluent guidelines added specific organics to the permit limits, specifically, the Organic Chemical, Plastics and Synthetic Fibers (OCPSF) and Pharmaceutical Effluent Guidelines. Aquatic toxicity, nutrients and volatile organic compounds, such as benzene and toluene, augmented the OCPSF regulations, and acetone, tetrahydrofuran and similar compounds appeared in pharmaceutical effluent guidelines. In recent years, the focus has changed to organics and trace organics, and to compounds of emerging concern or microconstituents, which include pharmaceuticals and personal care products and insecticides. Some constituents in these groups, such as estrogens and hormones are endocrine disruptors and are already being regulated in Europe and treated in activated sludge plants.

This book was developed to discuss these changes and thereby help operators, engineers and students evaluate alternatives and design and operate new activated sludge biological treatment plants and retrofit or upgrade existing plants. Novel approaches, technologies and tools and case studies are presented herein. This text can be used in workshops for students, treatment plant operators and practicing engineers, as well as for energy, environmental and utility managers and engineers in industry.

For many years, the authors taught the material presented in these pages at Manhattan College's Institute for Water Pollution Control (referred to as the Summer Institute), which originated in 1955. They have also conducted multiple workshops in the United States and abroad for design engineers, plant operators, industrial clients and students with case studies on both the trends in biological treatment technologies as well as the shift in focus from BOD and TSS to microconstituents such as active pharmaceutical ingredients (APIs). The results of these workshops, reflected in the present text, are intended to convey principles applicable on the job, as well as design guidance.

Activated sludge at 100 years old has proven to be, and will continue to be, a good technical solution to the treatment of industrial wastewater to meet the challenges of more stringent effluent limits for nutrients, specific organics and microconstituents as well as pretreatment for downstream water recycle and reuse technologies.

Acknowledgments

THE authors would like to acknowledge the assistance of several individuals who helped in the development of this book. Thanks to our wives, Agnes Eckenfelder and Maria Cleary, and Joe's three daughters, Danielle, Jacqueline and Tara for their love, encouragement and support to finish the book. Thanks to Kathleen F. Whartenby at KFW Business Support Services, Inc., who processed the content and kept me organized, plus Amanda Moore at HDR/HydroQual for her help. To my colleague, Gary Grey at HDR/HydroQual, for his technical review and guidance. To my colleague at HDR, John Schubert, who stepped up to write Chapter 8 on shale gas water management. To my co-workers and colleagues over the years and at HDR/HydroQual the last several years, and to my clients for the opportunity to work on some excellent teams and challenging and enjoyable projects. Thank you all.

CHAPTER 1

Introduction

THE activated sludge process roots mainly go back to studies at the Lawrence Experiment Station in Massachusetts during 1912 and 1913 by Clark and Gage [1]. These men experimented with aerated wastewater and growth of organisms cultivated in bottles and tanks partially filled with roofing slate. Their work led to Dr. G. Fowler of the University of Manchester, England to conduct experiments at the Manchester Sewage wastewater where Ardern and Lockett [2] carried out their research. Their presentation of April 3, 1914 [3] names the process activated sludge since it involved production of an activated mass of microorganisms compiled of aerobic stabilization of organic matter in wastewater and is now approaching 100 years old. These early works were the first to experiment with recycle of the biomass formed during the aeration period.

The principles of the activated sludge process presented in this book have been studied, better understood and modeled over the years, but the basic principles and fundamentals are still essentially the same. What has changed in both the industrial and municipal wastewater professions are the evolution of the numerous activated sludge process technologies and configurations, operating experience, effluent permit limits and advancements in process modeling tools. The tools available to design and troubleshoot the performance of activated sludge treatment technologies such as treatability and pilot studies and process modeling tools have also evolved. Figure 1.1 shows the process design and operational control parameters used for the basic activated sludge process which are discussed in more detail in Chapters 2 and 3 and include: equalization, pretreatment, activated sludge aeration basin and final clarifier.

Figure 1.2 shows the simple activated sludge process with an aeration basin and clarifier with the key nomenclature for the process variables shown in Figure 1.1.

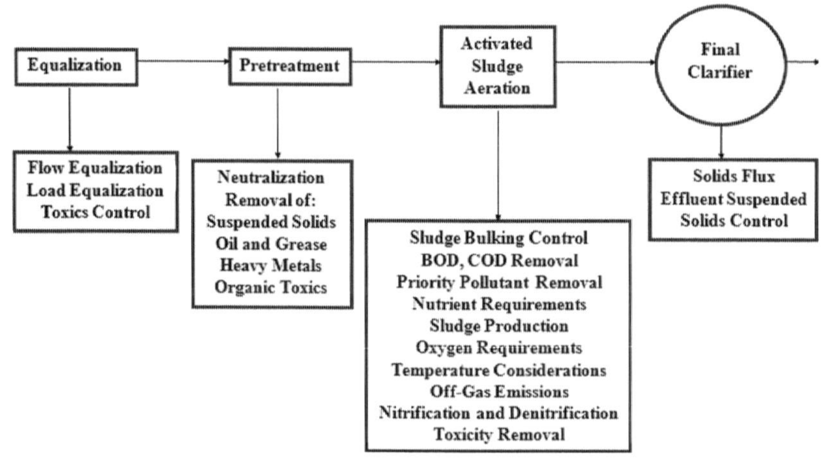

Figure 1.1. Process design and operational control variables for activated sludge treatment of industrial wastewaters [4].

The regulatory drivers have evolved from BOD and TSS removal for secondary treatment to nitrogen and phosphorus removal for advanced or tertiary treatment. Recent changes have included more stringent metals limits to removal of specific organics for the chemical and pharmaceutical industries. More recent trends in Europe and in the United States include the need to remove microconstituents such as pharmaceuticals and personal care products.

Activated Sludge Nomenclature

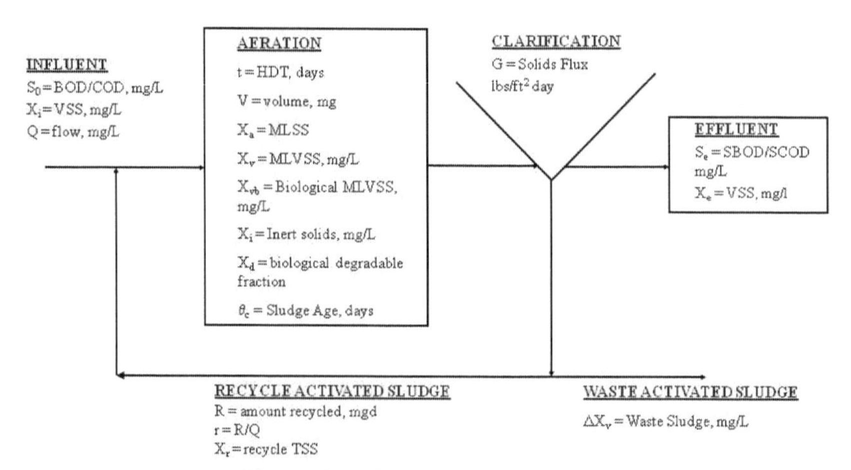

Figure 1.2. Activated sludge nomenclature.

Some of this is not regulatory driven, but by internal corporate standards in industry. Increasing emphasis on off-gas emissions has driven the need for covered aeration basins which can result in an excessively high temperature, resulting in floc dispersion and deterioration in effluent quality. In these cases, the influent wastewater may have to be cooled. While nitrification and denitrification are well defined for municipal wastewater, many organics present in industrial effluents are inhibitory to the nitrifying organisms. In these cases, design modifications are required to insure effective nitrification. Case studies of nitrification inhibition are presented herein in Chapters 5 and 7. In recent years, a major emphasis has been placed on aquatic toxicity. Many wastewaters, particularly those from the chemicals industry, require source treatment for toxicity reduction. In some cases, the addition of powdered activated carbon to the aeration basin has been employed to adsorb the non-degradable toxic organics.

The authors who have presented this subject matter at numerous workshops and at the Manhattan College Institute for Water Pollution Control decided that it was time to summarize these recent developments and trends over the last twenty years for engineers, operators and students. This book provides the reader a reference to understand the Principles of Activated Sludge (see Chapter 3) which have not changed over time. Wastewater Characterization and Pretreatment which are still critical to evaluating current performance of activated sludge treatment system and developing the design basis for plant upgrades are both discussed in Chapter 2. The trend to remove specific organics, nutrients, microconstituents (see Chapters 5 and 6) and apply activated sludge technology solutions (Chapter 4) are presented with case studies and design examples. The use of treatability study and process modeling tools which are becoming more valuable for both design and troubleshooting are presented in Chapter 7 along with examples.

Chapter 8 discusses shale gas water management from drilling operations and the wastewater characteristics and treatment options. The impacts of discharge to activated sludge plants and water reuse are discussed. Chapter 9 discusses water recycle and reuse treatment after the activated sludge process to help industry achieve sustainability goals for water reuse.

REFERENCES

1. Metcalf, L. and H.P. Eddy. 1930. *Sewerage and Sewage Disposal.* A Textbook, 2nd Edition. McGraw-Hill, New York.
2. *Wastewater Engineering Treatment and Reuse,* International Edition, Metcalf and Eddy. Fourth Edition, 2004.
3. Ardern, Edward and William Lockett. 1914. "Experiments on Oxidation of Sewage Without the Aid of Filters." Presentation to the Society of Chemical Industry at Grand Hotel in Manchester, England, April 3, 1914.
4. Eckenfelder, W.W. and Jack L. Musterman. 1995. *Activated Sludge Treatment of Industrial Wastewater.* Technomic.

CHAPTER 2

Wastewater Characterization and Pretreatment Needs

A key challenge which must be met on any industrial wastewater project is the development of the wastewater characterization, flow and pretreatment needs for the activated sludge process. The design basis should include a thorough understanding of the variability of both the flow and wastewater constituent concentrations. Changes in production, such as product campaigns in the pharmaceutical industry and future production increases or decreases over time should be included to build-in operational flexibility such as turn-up and turn-down ability. This flexibility in operation of the treatment plant can reduce the energy and annual O&M costs plus improve performance. More recent trends to reduce water footprint and to recycle treated water for utilities (e.g., scrubber and cooling tower) as discussed in Chapter 7, also need to be carefully considered in wastewater characterization and design basis development. Two examples of Design Basis are included at the end of Chapter 2. One design basis example included a wastewater characterization sampling program for a pharmaceutical plant which had very limited data, no treatment plant and specific effluent permit limits for acetone. The other design basis example was for a pulp and paper plant with a lot of data on waste streams and an existing treatment plant.

WASTEWATER CHARACTERIZATION

A wastewater characterization program always starts with a review of the existing data such as flow, TSS, COD, BOD, nitrogen and phosphorus and metals. Specific organics and micro-constituents are then added to the list as needed for a specific project. Data gaps are then typically identified and a sampling test plan is developed to both confirm the existing analytical data as well as fill in the data gaps. Detailed procedures for industrial wastewater surveys and wastewater sampling are presented elsewhere [1,2] and are not included in this book.

The wastewater characterization information is used to assess the potential biodegradability characteristics of the wastewater for activated sludge. The information is also used to determine if and what pretreatment processes are needed to remove constituents such as TSS, oil and grease, metals, ammonia, and inhibiting or toxic constituents to make the wastewater more suitable for activated sludge treatment. A treatability testing plan can then be developed, if needed, with appropriate pretreatment of the wastewater prior to activated sludge treatment. Treatability testing is discussed in Chapter 7. The wastewater characteristics needed for use in process models, such as Biowin and GPSX and also discussed in Chapter 7 should also be incorporated [2].

The BOD, COD and TOC concentrations are easy to measure and provide a more rapid indication of wastewater organics characteristics. The COD and TOC, however, both measure degradable and nondegradable organics, and, hence, adjustments in these measured values must be made to define those organics that are removable in the activated sludge process. Eckenfelder & Musteman (1995)[3] presents general guidelines that can be applied for the relationship between biodegradability and molecular structure. The BOD, TOC, and COD analyses indicate the "bulk" concentrations of degradable and nondegradable organics in the raw wastewater. These parameters determine the removal kinetics, oxygen requirements, aeration basin volume, and sludge production rates for the activated sludge process. In addition, raw wastewaters from certain industries must be further characterized to reliably define concentrations of compounds that may be inhibitory or toxic to the biomass, accumulated in the waste sludge, or specifically limited by air or liquid emission regulations. These include heavy metals, dissolved salts, sulfide, oil and grease, and certain volatile organic compounds (VOC) and semi-volatile organic compounds (SVOC) that are regulated as hazardous wastes. The design of the activated sludge process and the overall wastewater treatment system must consider these parameters by providing adequate pretreatment processes, off-gas capture and treatment, and sludge handling and disposal.

FLOW

A good design basis should include the projected flow, as well as the range in flow during the day as well as from day to day. It is very important to understand the manufacturing processes at a particular industry and develop an overall flow balance. The following is a checklist of questions to answer:

- Is the production and flow 5, 6 or 7 days per week?

- How many production shifts per day?
- Are the processes batch, continuous or both?
- What is the frequency and schedule of clean-in-place (CIP) operations?
- Is the process and sanitary wastewater segregated or combined?
- Is stormwater segregated?
- Does the flow change from month to month or seasonally?
- Are there plant shutdowns?
- Can high and low strength wastewater be segregated for separate treatment.

The design basis should include the following:

- Average daily flow rate
- Peak daily flow rate
- Peak hourly flow rate
- Minimum flow rates during the day
- Average constituent concentrations
- Peak constituent concentration
- Average mass loadings for key design parameters
- Peak mass loadings for key design parameters

A statistical summary and probability plots are recommended for both flow, concentrations and mass loadings of the key constituents.

EFFLUENT PERMIT LIMITS

The design basis should also include the target concentrations needed after treatment to meet the effluent permit limits. The effluent permit would typically have monthly average and daily maximum concentrations for a list of constituents. The treatment plant has to be designed to handle the variable flow and mass loadings and produce consistent effluent quality to ensure permit compliance. A summary of EPA effluent limitation guidelines for various industrial categories are presented elsewhere [2].

PRETREATMENT

Efficient operation of the activated sludge process requires control of pollutants such as toxic substances, incompatible pollutants such as acids/bases, oil and grease, and fluctuations in hydraulic and organic loads through a variety of pretreatment technologies. Table 2.1 shows guidelines and concentrations of pollutants and pretreatment technologies. Figure 2.1 shows a general example of pretreatment technologies utilized prior to activated sludge [3].

TABLE 2.1. Concentration of Pollutants that Make Prebiological Treatment Desirable [3].

Pollutant or System Condition	Limiting Concentration	Kind of Pretreatment
Suspended solids	< 50 to 125 mg/L	Sedimentation flotation, lagooning
Oil or grease	< 35 to 50 mg/L	Skimming tank or separator
Toxic ions		
Pb	≤ 0.1 mg/L	
Cu + Ni + Cn	≤ 1 mg/L	Precipitation or ion exchange
Cr^{+b} + Zn	≤ 3 mg/L	
Cr^{+3}	≤ 10 mg/L	
pH	6 to 9	Neutralization
Alkalinity	0.5 lb alkalinity as $CaCO_3$/lb BOD removed	Neutralization for excessive alkalinity
Acidity	Free mineral acidity	Neutralization
Organic load variation	< 2:0	Equalization
Sulfides	< 100 mg/L	Precipitation or stripping with recovery
Phenols	< 70 to 300 mg/L	Extraction, adsorption, internal dilution
Ammonia	< 500 mg/L (as N)	Dilution, ion exchange, pH adjustment and stripping
Dissolved salts	<10 to 16 g/L	Dilution, ion exchange
Temperature	13 to 38°C in reactor	Cooling, steam addition

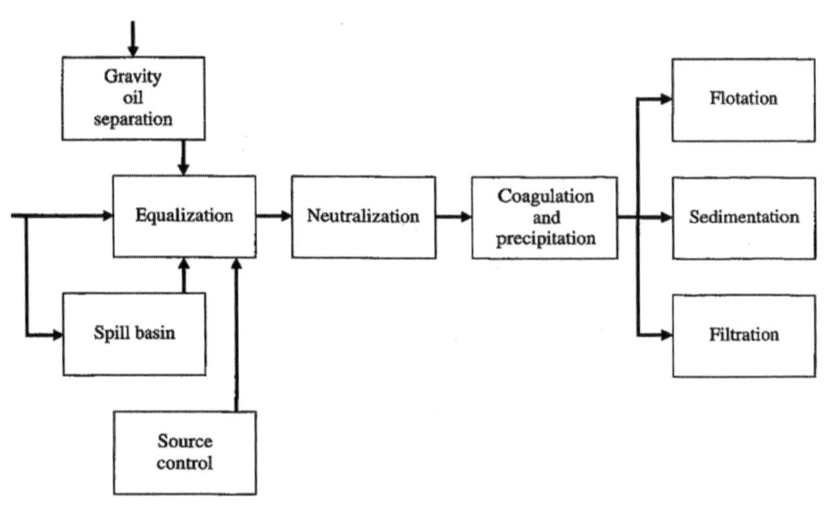

Figure 2.1. Alternative pretreatment technologies prior to activated sludge process [3].

Equalization

Most industrial wastewater dischargers fluctuate in flow rate and/or pollutant mass loading with time due to variability in batch processing and campaign production changes. Equalization is almost always required to dampen these fluctuations and maintain stable operation of the pre-treatment and activated sludge process. The equalization system should be completely mixed and can be operated in either a constant volume (variable outflow) or variable volume (constant outflow) mode. If the wastewater flow rate remains reasonably constant with time, a constant volume basin will provide adequate load equalization. By contrast, a plant employing batch production processes with rapid changes in both flow and pollutant load should employ a variable volume basin with a constant volumetric withdrawal rate or be fed on a batch basis each day [7]. Design guidelines and procedures are presented elsewhere [1] for equivalization and off-line storage. It is typical to have at least 12 to 24 hours equalization basin detention time.

In cases of readily degradable wastewaters, aeration should be provided in order to avoid septic conditions. In most cases, aeration at 20 hp/MG (million gallons) basin volume should be adequate, but actual mixing power requirements should be based on the settling characteristics of the wastewater's suspended solids.

Spill Control

In industries subject to spill events and/or periodic shock loads, a spill basin should be provided to divert the influent wastewater flow when the concentration exceeds a predetermined value. Parameters requiring spill basin control are wastewater strength (usually defined as TOC or TOD), TDS, temperature, and toxic organics and metals. The characteristics of the potential spill material must be suitable for continuous on-line analysis so that an instrument signal can divert the flow to the off-line basin. Reliable on-line monitoring of dissolved salts (conductivity), TOC, TOD, temperature, biomass respiration rate, and purgeable organic carbon (POC) have been provided. A flow schematic of an off-line spill basin is shown in Figure 2.2.

Oil and Grease Removal

High concentrations of oil, such as those found in petroleum refining wastewater or food industry (e.g., chicken and bakery plants), should be removed by gravity separation in an API or corrugated plate separator or dissolved air flotation (DAF) system prior to activated sludge treatment. In many cases, the refinery API basin is followed by a dissolved air flotation unit (DAF), in order to achieve levels of 15 to 20 mg/L that are compatible with the biological

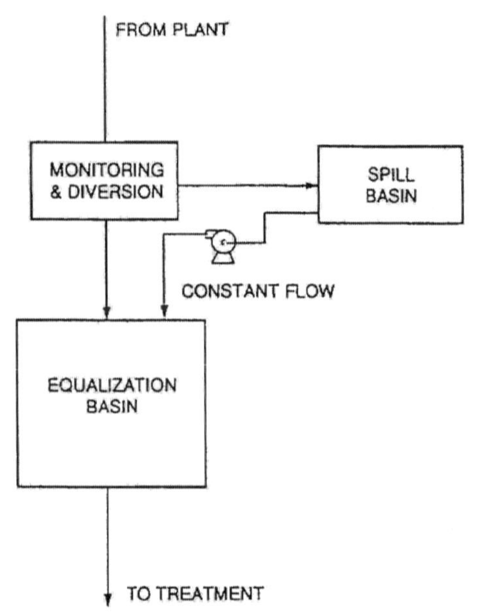

Figure 2.2. Flow schematic for off-line control of spills and shock loads [3].

treatment process and secondary clarification requirements. Food industries typically have DAF treatment with chemical treatment.

Heavy Metals

The presence of heavy metals in the raw wastewater poses several problems. First, they contribute to effluent aquatic toxicity; second, they have restrictive permit limits; and third, bioconcentration effects can result in high metals levels in the stabilized sludge and thereby limit the options for ultimate solids disposal. Metals and cyanide removal through the activated sludge process have been documented by many investigators [4] and can range from 40 to 80%. The variability of metals and cyanide removal through the activated sludge process is related to the following removal mechanisms:

- Entrapment of precipitated metals in the sludge floc matrix. Parameters that affect operating pH and floc size and character will also affect metals removal.
- Bacterial extracellular polymer binding of soluble metals. Extracellular polymer production is a function of sludge age or organic loading rate.
- Accumulation of soluble metals by the cell.
- Biodegradation and volatilization of cyanide as affected by the operating SRT and aeration rate.

TABLE 2.2. Heavy Metal Removal in an Activated Sludge Process Treating Petroleum Refinery Wastewater [3].

	Activated Sludge Plant	
Heavy Metal	**Influent (mg/L)**	**Effluent (mg/L)**
Cr	2.2	0.9
Cu	0.5	0.1
Zn	0.7	0.4

Metals removal in the activated sludge process from a petroleum refinery wastewater is shown in Table 2.2. As the solids retention time is increased, metal accumulation on the sludge will also increase, as shown for accumulation of copper in Figure 2.3. Metals should be removed prior to the activated sludge process. Technologies for their removal and attainable effluent levels are listed in Table 2.3.

pH Neutralization

The activated sludge process operates most effectively over a pH range of 6.5 to 8.5, and neutralization may be required for wastewaters that are outside of this pH range. There are exceptions, however, in which highly alkaline or

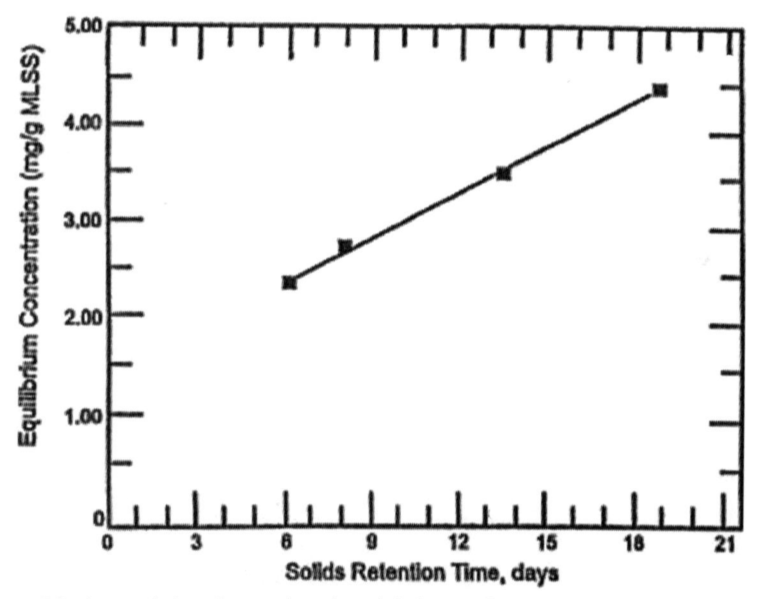

Figure 2.3. Accumulation of copper in activated sludge as a function of solids retention time [3].

TABLE 2.3. Heavy Metals Removal Technologies and Achievable Concentrations [3].

Metal	Technology	Achievable Concentration (mg/L)
Arsenic	Sulfide precipitation (pH 6 to 7) Carbon adsorption Co-precipitation (ferric hydroxide)	0.05 0.06 0.005
Barium	Sulfate precipitation	0.5
Cadmium	Hydroxide precipitation (pH 10 to 11) Co-precipitation (ferric hydroxide) Sulfide precipitation (pH 6 to 7)	0.05 0.05 0.008
Copper	Hydroxide precipitation Sulfide precipitation (pH 8.5)	0.02 0.01
Lead	Carbonate precipitation (pH 9 to 9.5) Hydroxide precipitation (pH 11.5)	0.01 0.02
Mercury	Sulfide precipitation Co-precipitation (alum) Co-precipitation (ferric hydroxide)	0.01 0.001 0.0005
Nickel	Hydroxide precipitation (pH 10 to 11)	0.12
Zinc	Hydroxide precipitation (pH 10)	0.1

acidic wastewaters do not require pH adjustments for effective treatment by activated sludge. If a complete mix activated sludge process is used, hydroxyl ions (OH^-) will react with the carbon dioxide (CO_2) produced by microbial respiration yielding bicarbonate ion (HCO_3^-), which will buffer the system at a pH near 8.0. As a general rule, 0.5 lb of hydroxide alkalinity (as calcium carbonate) will be neutralized by 1 lb of BOD removed in the process. High-strength textile mill wastewaters, having a pH of 10 or greater, entering a complete mix biological process were neutralized to a pH near 8.0 by microbial respiration and CO_2 production. Similar neutralization effects occur in high-pH wastewaters that have been lime treated to pH 10 or 11 for phosphorus and/or metals removal. It should be noted, however, that the amount of hydroxide alkalinity, rather than the actual pH of the influent wastewater, determines the degree of neutralization required.

In like manner, organic acids will be biologically oxidized to CO_2 and water. The CO_2 will be air-stripped from the process to its equilibrium concentration, thus reducing the acidity. Wastewaters from the synthetic fibers and organic chemicals industries containing acetic acid and having a pH less than 4.0 have been successfully treated in a complete mix activated sludge process without preneutralization.

There are also cases in which the biological reactions generate acidity or alkalinity. Nitrification, which is discussed in Chapter 5, generates acidity and destroys alkalinity, whereas denitrification produces alkalinity. In many cases, supplemental alkalinity must be added to offset the effect of nitrification and maintain an optimal pH in the treatment process. Biological oxidation of sulfonates yields sulfuric acid (H_2SO_4). The wastewater from one plant containing sulfonates required pH adjustment to 11.5 prior to entering the aeration basin in order to maintain optimal mixed liquor pH. Wastewaters containing high concentrations of salts of weak organic acids, e.g., sodium acetate, will produce alkalinity in the form of carbonates and cause the mixed liquor pH to increase to a range of 8.5 to 9.2. Similarly, wastewaters that contain high concentrations of organically based nitrogen will cause mixed liquor pH values of 9.0 to 9.5 due to biohydrolysis of the amine to ammonium (NH_4^+). These operating conditions exceed the recommended mixed liquor pH range and may require acid addition directly to the aeration basin.

Mineral acidity will require pH adjustment prior to the biological process. In cases of variable influent pH, neutralization will usually follow or take place in the equalization process in order to utilize the equalization basin contents. In cases of highly acidic wastewaters, neutralization will usually precede equalization in order to minimize corrosion. A two-stage equalization process is usually necessary if large changes in pH are required and/or if a narrow effluent pH range must be maintained. The number of stages required for pH control depends primarily on the shape of the titration curves.

Toxic Substances and Off-Gas Control

Toxic organic compounds in wastewaters should be removed by pretreatment prior to the activated sludge process. This includes organics that are toxic to aquatic life, as measured by bioassay methods, as well as organics that are inhibitory to the biomass in the activated sludge process. Technologies for pretreatment of toxic wastewaters include: precipitation and solids removal or ion exchange for heavy metals; chemical oxidation; wet air oxidation steam stripping for organics; and air or steam stripping for volatile organics. It should be noted that, in many cases, the primary objective of the pretreatment process is detoxification and enhanced biodegradability.

Restrictions on air emissions may require prestripping and off-gas capture prior to activated sludge treatment of wastewaters containing high concentrations of volatile organic and inorganic substances (VOS). If the VOS are not controlled and treated at the stripper, they will be volatilized at the aeration basin where off-gas capture and control is more difficult and costly.

Ammonia-Nitrogen

Ammonia-Nitrogen is inhibiting to the biomass at concentrations of about 1,500 mg/L. If the biological system is designed for just BOD removal, pretreatment of the ammonia-N would typically be compared to biological nitrification. Pretreatment for ammonia-N can be done using air or steam stripping or vacuum distillation. The ammonia removal pre-treatment can be applied whenever ammonia concentrations exceed 1,000 mg/L such as in landfill leachate, food waste organics to energy plants and industrial wastewater treatment plants. Air stripping and vacuum distillation treatment has been used for ammonia removal.

Organics

Anaerobic biological treatment which is described in detail in Chapter 3, is typically used for pre-treatment of high strength wastewater with COD over 3,000 mg/L prior to activated sludge. It is used primarily in the food and beverage, rum distillery and occasionally in chemical and pharmaceutical industry. It is typically used to treat readily biodegradable wastewater to reduce the size, foot print, energy and sludge disposal requirements of downstream activated sludge plants and also produce biogas for energy recovery and are in boilers or to make electricity. Air and steam stripping technologies have typically been used in the chemical and pharmaceutical industry to remove volatile organics and solvents for subsequent recovery or disposal.

Example 2.1—Design Basis for Pharmaceutical Plant

This design basis example was for a pharmaceutical plant with high strength COD and ammonia wastewater and an existing nitrification and denitrification biological treatment plant which needed additional capacity to handle the increased flow and nitrogen loads. Table 2.4 shows the design basis for the 2009 expansion versus the original 2002 design basis. The percent increase in flow, COD and nitrogen were 76, 43 and 25 percent, respectively. The key effluent permit limits for discharge to the POTW was an ammonia-nitrogen limit of 17.3 mg/L average and 50 mg/L peak. This activated sludge plant, with nitrogen removal, typically achieved a BOD less than 20 mg/L and an ammonia-nitrogen less than 5 mg/L. These design basis values were developed from statistical analysis of the daily flow and loads over a period of two years plus future production increases from the client.

Example 2.2—Design Basis for Pulp and Paper Plant

This design basis was developed for a pulp and paper plant in South Amer-

TABLE 2.4. Design Basis Flows and Wasteloads for Pharmaceutical Plant.

	Design Basis (2009)	Design Basis (2002)	% Increase
Average flow gpd	300,000	170,000	76
Peak flow, gpd	420,000	180,000	230
Average COD, lbs/day	23,250	16,275[a]	43
Peak COD, lbs/day	29,600	20,720[a]	43
Average BOD, lbs/day	13,780	9,645	43
Peak BOD, lbs/day	17,545	12,279	43
Average TKN, lbs/day	2,500	2,012	25
Peak TKN, lbs/day	2,900	2,334	25
Average TSS, lbs/day	1,500	87	
Peak TKN, lbs/day		99	

[a]Calculated based on BOD/COD rate of 0.65.

ica. Wastewater flow projections were evaluated with the client based on both expected production increases and in-plant water reduction projects that were in the process of being implemented. The wastewater concentrations of COD, BOD and TSS from the eight individual wastewater streams were analyzed to develop a flow and material balance. These streams combine and flow to the primary clarifier. The primary clarifier effluent flows to the aerated stabilization lagoon. Table 2.5 shows the flow and loads for the individual wastestreams.

Table 2.6 shows the design basis total flow and loads used for segregation of the high strength wastewater (desmedulado which is equivalent to depithing in U.S.) for separate anaerobic treatment to produce biogas for energy recovery and to reducc the BOD load and sludge production in the aerated lagoon. The other streams receive primary clarification and biological treatment in the ASB. An activated sludge treatment plant is planned to replace the ASBs which were failing on TSS removal due to build-up of sludge.

TABLE 2.5. Flow and Wasteloads for Pulp and Paper Plant.

Plant #1: Yumbo	Flow (gpm)	BOD_5 (kg/day)	TSS (kg/day)	COD (kg/day)
Desmedulado	1,256	14,362	19,981	36,927
Pulpa	455	2,036	6,191	7,098
Blanqueo	991.5	3,505.5	1,575.1	9,121.0
Inorganicos	502.7	729.2	2,621.2	4,788.6
Maquina 1	1,604.2	1,653.9	2,621.2	4,788.6
Maquina 2	530.7	835.1	2,007.0	2,267.5
Maquina 3	1,860.4	2,133.4	5,841.9	9,596.9
Pozo	204.8	768.1	5,347.2	3,447.3

TABLE 2.6. Design Basis Flow and Wasteloads for Pulp and Paper Plant.

Production	Water Usage (m^3/ton paper)	Flow (GPM)	BOD_5 (kg/day)	TSS (kg/day)	COD (kg/day)
Existing Data (117,000 ton/yr production)	126	7,405	26,023	46,267	77,605
Projected Data (125,000 ton/yr production)	100	6,191	34,120	58,371	100,019
Production	**Water Usage (m^3/ton paper)**	**Flow (GPM)**	**BOD_5 (mg/L)**	**TSS (mg/L)**	**COD (mg/L)**
Existing Data (117,000 ton/yr production)	126	10.7	644	1,145	1,920
Projected Data (125,000 ton/yr production)	100	8.92	1,010	1,727	2,959

REFERENCES

1. *Wastewater Engineering Treatment and Reuse,* International Edition. Metcalf and Eddy. Fourth Edition 2004.
2. *Industrial Wastewater Management, Treatment and Disposal,* Third Edition, Manual of Practice No. FC-3. Water Environment Federation (WEF). 2008.
3. Eckenfelder, W.W. and Jack L. Musterman. 1995. *Activated Sludge Treatment of Industrial Wastewater.* Technomic.
4. Lester, J.C. 1987. *Heavy Metals in Wastewater and Sludge Processes—Volume II, Treatment and Disposal.* CRC Press.

CHAPTER 3

Activated Sludge Process Principles

A greater understanding of activated sludge process principles has evolved over time with experience gained through the design and operation of hundreds of plants and the extensive research and development. There are a multitude of books on the activated sludge process including those by Eckenfelder [1,2] and others [3,4]. This chapter discusses the basic principles of the activated sludge process which are key for both design of new and upgrades of existing industrial wastewater plants plus troubleshooting problems at plants. These fundamental principles, which the authors have taught at the Manhattan College Summer Institute program for many years, are still relevant today for design and troubleshooting the numerous activated sludge technologies which are discussed in Chapter 4. The following basic principles are discussed in this chapter.

- Food to Microorganism Ratio
- Sludge Age
- Kinetics
- Temperature
- Sludge Production
- Oxygen Utilization
- Nutrient Requirements
- Acclimation
- Bioinhibition
- Control of Activated Sludge Quality
- Effluent Suspended Solids Control
- Oxygen Transfer

Definitions are shown below for both activated sludge and activated sludge process:

- *Activated Sludge*—"Sludge floc produced in a raw or settled sewage by the growth of zoogleal bacteria and other organisms in the presence of dissolved oxygen, and accumulated in sufficient concentration by returning floc previously formed."
- *Activated Sludge Process*—"A biological sewage treatment process in which a mixture of sewage and activated sludge is agitated and aerated. The activated sludge is subsequently separated from the treated sewage (mixed liquor) by sedimentation and wasted or returned to the process as needed.

MICROBIOLOGY

Activated sludge flocs contain bacteria, organics, and inorganics. Floc size ranges from < 1 to 1000 μm or more, and viable bacteria make up approximately 5 to 20% [3]. Although it is assumed that there is enough oxygen throughout the aeration tank, the oxygen distribution within the activated sludge is subject to mass-transfer limitations. The surface of the floc is aerobic, but an anoxic zone exists inside and there is a small anaerobic zone at the center.

Activated sludge floc typically includes [3]:

- Bacteria (e.g., *Zooglea, Pseudomonas, Flavobacterium, Alcaligenes, Bacillus, Achromobacter, Corynebacterium, Comomonas, Brevibacterium,* and *Acinetobacter*);
- Filamentous organisms (e.g., *Sphaerotilus, Beggiatoa,* and *Vitreoscilla*);
- Autotrophic bacteria (e.g., nitrifiers [*Nitrosomona*s and *Nitrobacter*] and phototrophic bacteria [*Rhodospirrilaceae*]);
- Protozoa (e.g., ciliates, flagellates, and *Rhizopoda*); and
- Rotifers (e.g., *Bdelloidea, Monogononta, Lecane* sp., *Notommato* sp., *Philodina* sp., and *Habrotrocha* sp.)

In the activated sludge process, organic matter is removed from solution by biological metabolism, oxygen is consumed by the organisms, and new cell mass is synthesized. The organisms also undergo progressive auto-oxidation (endogenous decay) of their cellular mass. Figure 3.1 depicts the course of bio-oxidation of an organic wastewater. When a wastewater is mixed with acclimated biological sludge, there may be an immediate "sorption" of readily degradable organics. These organics are stored within the cell for subsequent oxidation. This phenomenon is called biosorption. As aeration proceeds, removal of the remaining organics occurs. The oxygen uptake rate is initially high as the sorbed organics are degraded and then decreases as the residual substrate decreases. Cellular synthesis occurs in proportion to the organic re-

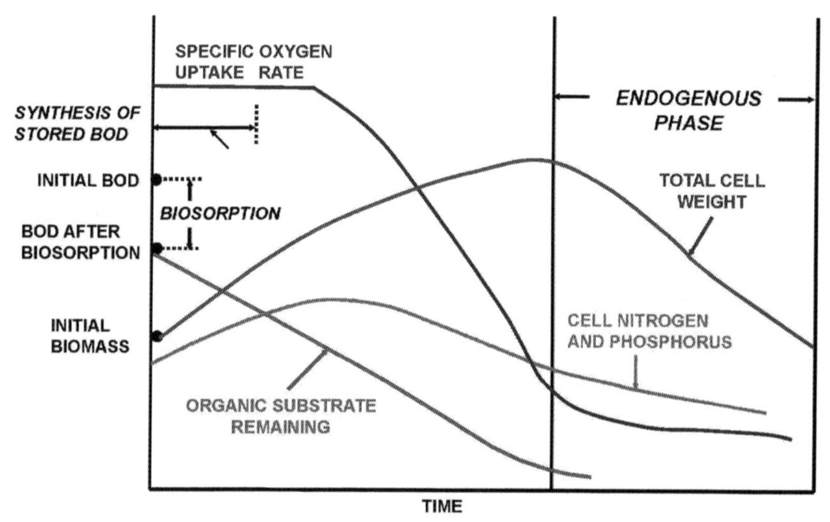

Figure 3.1. Reactions occurring during bio-oxidation [1].

moval. Nitrogen and phosphorus are taken up for cellular synthesis. When the available substrate is exhausted, continued aeration results in oxidation of the biomass through endogenous respiration. These phenomena are described by Equations (3.1) and (3.2):

Synthesis of New Cells using BOD, Nutrients and Oxygen

$$\text{Organics} + aO_2 + N + P \frac{\text{cells}}{K} \rightarrow a(\text{New Cells}) + CO_2 + H_2O + \text{SMP} \quad (3.1)$$

$$a = \frac{\text{lbs cells (VSS) synthesize } d}{\text{lbs BOD or COD removed}} \quad a' = \frac{\text{lbs } O_2 \text{ consumed}}{\text{lbs BOD or COD removed}}$$

$$K = \text{BOD removal rate coefficient, } d^{-1}$$

Endogenous Respiration—Oxidation of Degradable Cellular Mass

$$b\text{Cells} + b'O_2 \rightarrow CO_2 + H_2O + N + P + \text{Nondegradable cellular residue} + \text{SMP} \quad (3.2)$$

$$b = \text{endogenous rate, fraction per day degradable biomass oxidized, } d^{-1}$$

$$b' = 1.4 \times b, d^{-1}$$

In Equation (3.1), a' is the oxygen equivalent of the fraction of organic matter removed that is oxidized to end products (i.e., CO_2 and H_2O). Term a is the fraction of organic matter removed that is synthesized into biomass, and K is a temperature-dependent reaction rate coefficient that is related to the biodegradability of the specific wastewater. In Equation 3.2, the coefficient b is the fraction per day of degradable biomass that is endogenously oxidized, and b' is the oxygen required to support the endogenous decay.

A small portion of the organics removed for synthesis [Equation (3.1)] and endogenous metabolism [Equation (3.2)] remains as nondegradable cellular residue and SMP_{nd}. The nondegradable cell residue is approximately 20 percent of the volatile suspended solids generated in Equation (3.1). The SMP_{nd} varies from 2 to 10 percent of the influent degradable SCOD and depends on the organic composition of the raw wastewater and sludge age.

Since all but a small portion of the organics removed are either oxidized to end products or synthesized to biomass, the coefficients a and a' can be combined on a total oxygen demand basis using either COD or the adjusted ultimate BOD value (i.e., $BOD_u/0.92$). On a COD basis, their relationship is

$$a_{COD} + a'_{COD} = 1 \tag{3.3}$$

Their actual sum, however, will always be less than unity due to the formation of SMP_{nd}. Since the biomass is usually expressed as volatile suspended solids (VSS), it is convenient to express a in VSS units. On the average, it takes 1.4 g O_2 to oxidize 1 g of biomass expressed as VSS. Therefore, Equation (3.3) can be modified to

$$1.4(a'_{VSS}) + a_{COD} = 1 \tag{3.3a}$$

where,

a = mg VSS produced per mg of COD (or $BOD_u/0.92$) removed
a' = mg O_2 consumed per mg of COD (or $BOD_u/0.92$) oxidized to end products

These concepts are illustrated in the material balance for COD in Figure 3.2.

In order to design an activated sludge process, the coefficients a, a', b, b', and K must be determined (or assumed) for the specific wastewater-sludge mixture. This permits material balances to be developed around the process for substrate removal, oxygen requirements, and biomass production. Values of these parameters are available for common industrial wastewaters such as those from petroleum refining, food processing, and pulp and paper manufacturing. Wastewaters from the organic chemicals and pharmaceutical industries, however, require laboratory or pilot plant studies to define these parameters.

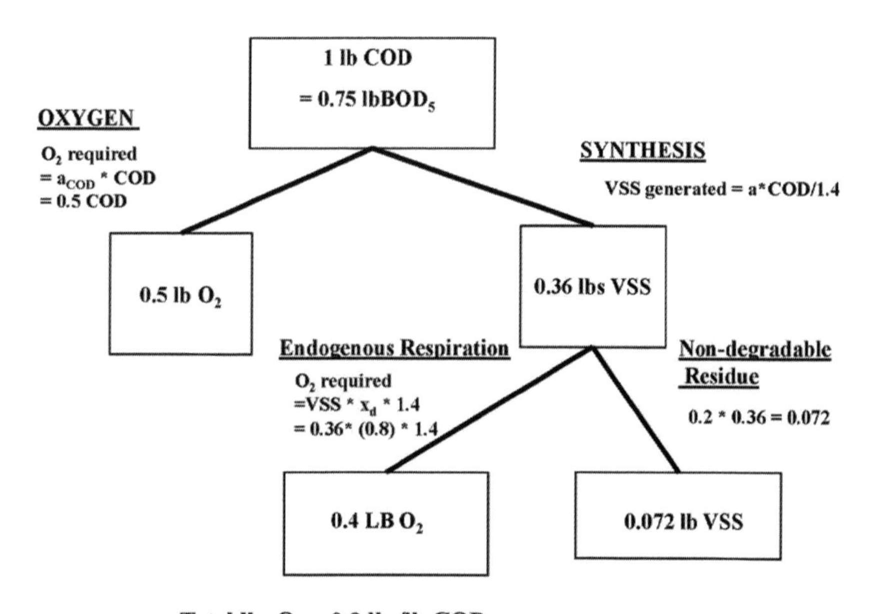

Figure 3.2. Material balance calculations for oxygen and VSS [4].

FOOD TO MICROORGANISM RATIO

The design basis and operating criteria for an activated sludge process are expressed in terms of the food to microorganism ratio (F/M) and the sludge age (θ_c) or solids retention (SRT). The F/M is the mass of organics in the influent (i.e., BOD or COD) divided by the MLVSS in the system.

$$F/M = \frac{S_o}{X_v t} \tag{3.4}$$

where,

F/M = food to biomass ratio, mg/mg-day
S_o = BOD or degradable COD in the influent, mg/L
X_v = biomass under aeration, mg VSS/L
t = hydraulic retention time, days

SLUDGE AGE

The sludge age or solids retention time (SRT) or mean cell residue time (MCRT) are all the same and is expressed as

$$\text{SRT} = \theta_c = \frac{X_v T}{\Delta X_v} \tag{3.5}$$

where,

θ_c = sludge age, days
ΔX_v = volatile suspended solids produced and wasted per day, mg/L
$X_v t$ = volatile suspended solids in reactor, mg-day/L

The *F/M* and sludge age relationship is as follows:

$$\frac{1}{\theta_c} = a\frac{F}{M} - bX_d \tag{3.6}$$

where,

a = lbs cells VSS synthesized/lb BOD removed
b = Endogenous rate, fraction pre day degradable biomass oxidized, d^{-1}
X_d = Degradable fraction of the VSS

Figure 3.3 shows the optimum range of both *F/M* sludge age where BOD removal and sludge settleability is achieved with a well flocculated activated sludge which can be visually observed using a microscope or in a settling column or settleometer test. .

PLUG FLOW AND COMPLETE MIX

Figure 3.4 shows the types of activated sludge processes: plug flow, complete mix and selector-complete mix. The plug flow process has been used for years for treatment of municipal wastewater while complete mix has been used mostly for industrial wastewater. The selector was added later to promote the growth of floc former organisms over filaments. Chapter 4 presents the evolution of activated sludge process configurations and technologies from the original plug flow and complete mixed processes.

KINETICS

Most industrial wastewaters contain multi-component mixtures of various types of organics (carbohydrates, protein, fatty acids, organic complexes, etc.). These organics can be broadly classified as "readily degradable" (e.g., dairy, food processing, pulp and paper) or "biorefractory" (e.g., tannery, agricultural chemicals, and certain pharmaceuticals). The removal characteristics of these wastewaters by activated sludge are depicted in Figure 3.5.

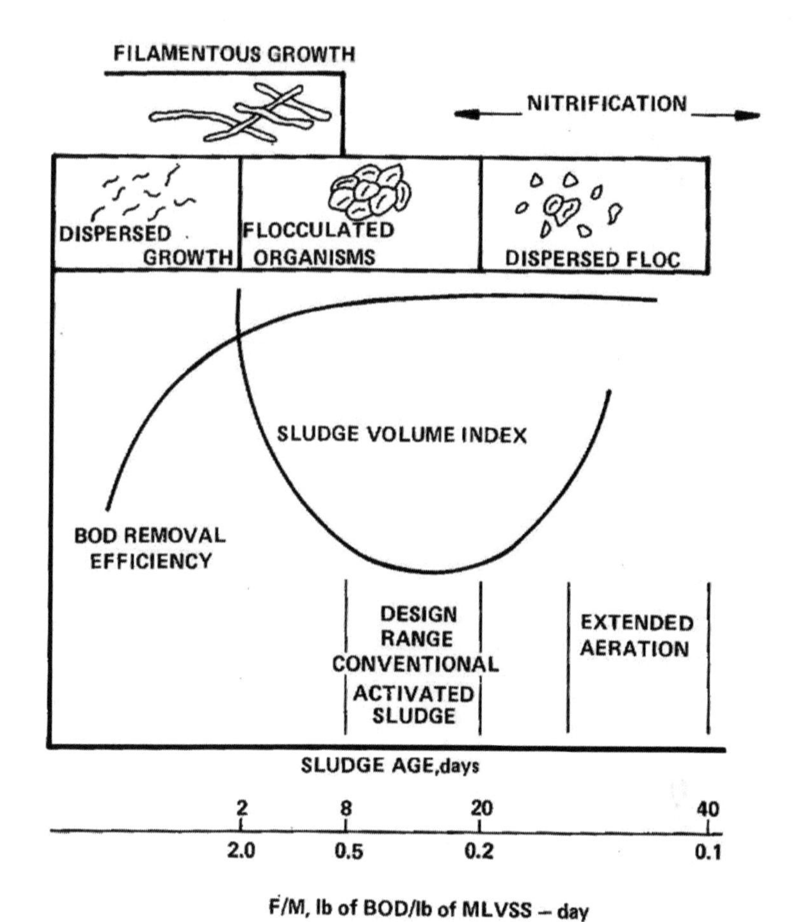

Figure 3.3. Flocculation and settling characteristics of activated sludge as related to organic loading.

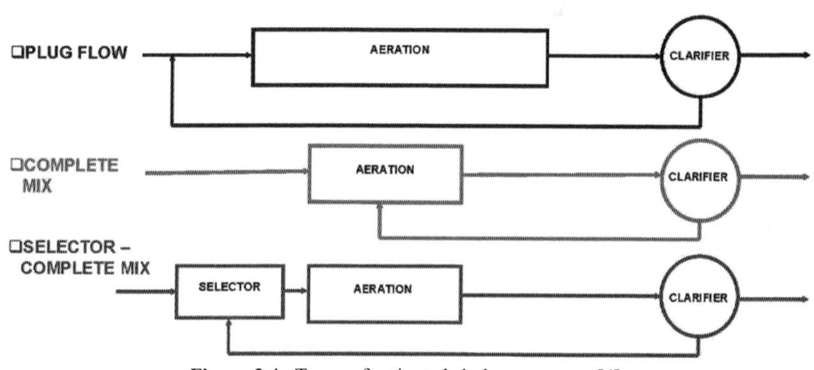

Figure 3.4. Types of activated sludge processes [4].

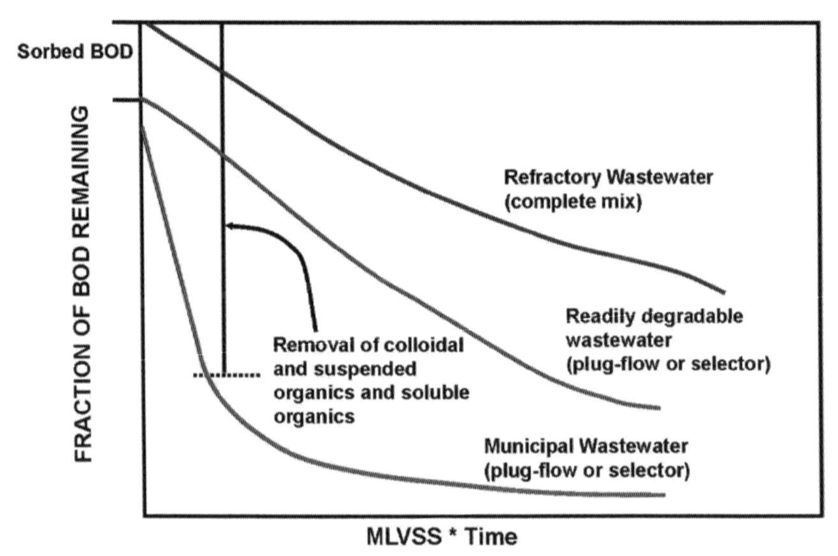

Figure 3.5. Organic removal characteristics by activated sludge [4].

In the case of a readily degradable wastewater, there is a rapid "sorption" of organics by the biological floc immediately on contact of the wastewater with the return sludge. The amount of biosorption depends on the activity of the biomass (i.e., the SRT) and the characteristics of the wastewater. For poorly degradable wastewaters or those containing high concentrations of amino acids, protein, or acetate, there may be little or no sorption. For comparative purposes, the activated sludge removal characteristics of typical domestic sewage are also shown in Figure 3.4. Domestic sewage contains approximately one-third suspended organics, one-third colloidal organics, and one-third soluble organics. On contact with activated sludge, the suspended organics are removed by being enmeshed in the biological floc. The colloidal organics are partially adsorbed and entrapped by the floc and a portion of the soluble organics are "sorbed" by the floc. The result is up to 85 percent removal of the total COD after 10 to 15 min of wastewater-sludge contact. The remaining degradable soluble organic fraction undergoes biological reaction in accordance with the kinetics of the system. The biosorption phenomenon was the basis for development of the contact-stabilization process for treatment of food processing wastewaters.

The use of a complete mix activated sludge process for treatment of a readily degradable wastewater will result in filamentous bulking. In these cases, a plug-flow regime or a biological selector located ahead of the complete mix basin should be used. Wastewaters that contain poorly degraded substrates generally do not stimulate filamentous growth and can be treated in a complete mix flow configuration.

The extent of the substrate (expressed as BOD or COD) biosorption can be related to the "floc load" (FL) in the biological selector [5]. Floc load is defined as the mg COD (or BOD) applied/g VSS and is expressed by Equation (3.6) for a given selector hydraulic retention time.

$$\mathrm{FL} = \frac{Q_o f_{sot} S_o}{Q_R X_d f_b X_v} \tag{3.7}$$

where,

Q_o = wastewater flow rate, MGD
Q_R = return sludge flow rate, MGD
S_o = soluble substrate, mg/L
f_{sor} = fraction of So that can be biosorbed
X_d = degradable fraction of VSS
f_b = fraction of VSS that is biomass
X_v = VSS, mg/L

Sorption versus floc load relationships obtained for several wastewaters are shown in Figure 3.6. The sorption efficiency is related to the active mass of the sludge, which decreases as the *F/M* decreases. A lower *F/M* (or higher SRT) implies a lower degradable fraction of biomass and hence a lower active mass.

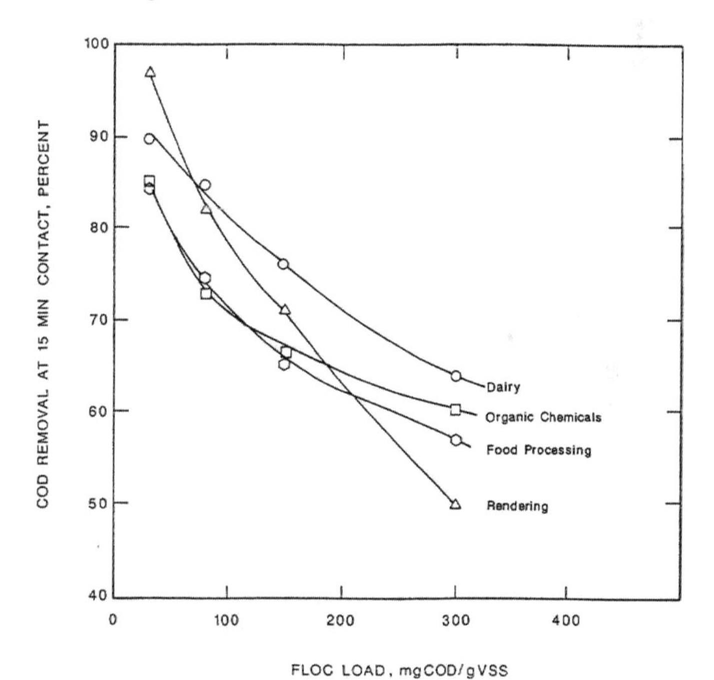

Figure 3.6. COD removal by biosorption vs. floc load for several industrial wastewaters [4].

Plug-Flow Kinetics

It has been shown that, under acclimated biomass conditions, most individual organic compounds are removed simultaneously at a linear rate, i.e., a zero order reaction, to low residual levels. Depending on the compound, the rates will be different, as shown in Figure 3.7. Consider the case of the simultaneous removal of three organics (A, B, and C), as illustrated in Figure 3.7. When the organics are measured as COD, BOD, or TOC, a linear rate will exist until time, t, when component A is substantially removed, leaving only compounds B and C. A reduced removal rate will then exist until time t_2, when component B is substantially removed. This results in a further reduction in rate for removal of only component C. Since most wastewaters contain numerous organic constituents, the organic removal with time will plot as a curve, as shown in Figure 3.7(b). When plotted as a semi-logarithmic function, a linear relationship results, as shown in Figure 3.7(c). Plug-flow BOD removal kinetics for a pulp and paper mill wastewater are shown in Figure 3.8. Batch or plug-flow kinetics can be defined by Equation (3.8):

$$\frac{S_t}{S_i} = e - K_b X_d X_v t / S_i \tag{3.8}$$

where,

S_i = substrate concentration remaining after biosorption, mg/L
S_t = substrate concentration remaining at t, mg/L
K_b = kinetic coefficient, day^{-1}
X_v = VSS in reactor, mg/L
t = time, days

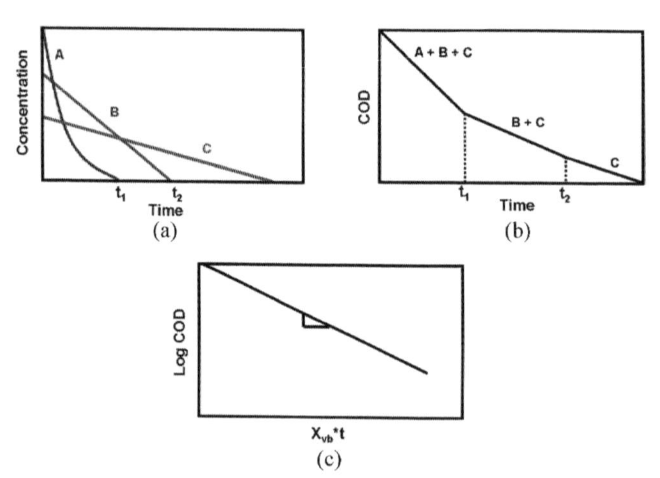

Figure 3.7. Schematic representation of multi-component substrate removal [4].

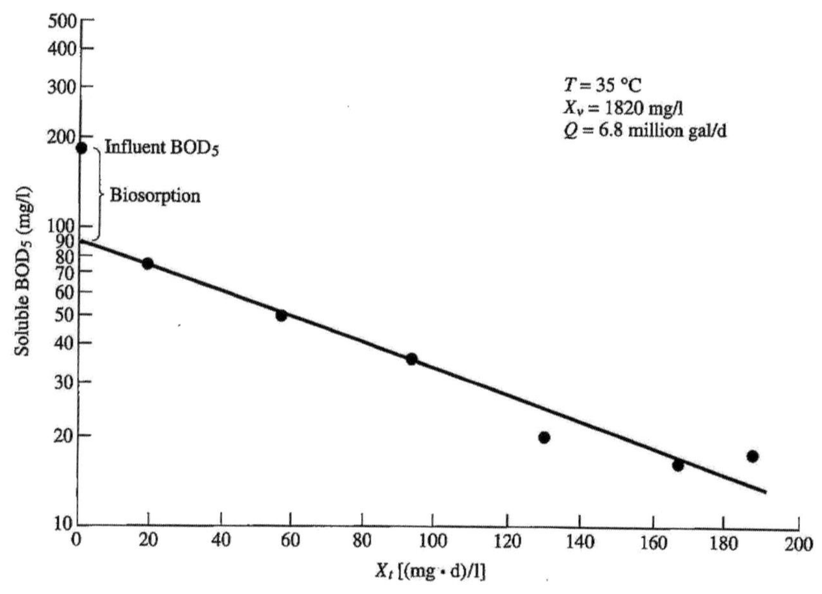

Figure 3.8. Plug flow BOD removal kinetics for a bleached Kraft Pulp and Paper wastewater [4].

Complete Mix Kinetics

In a complete mix reactor, the overall substrate removal rate decreases as the concentration of organics remaining in the reactor decreases. This is because the more readily degradable organics are removed first, resulting in a mixture of substrates and metabolites that are progressively more difficult to degrade. The relationship defining substrate removal in the complete mix-activated sludge process is presented as Equation (3.9):

$$\frac{S_O - S_e}{X_v t} = K \frac{S_e}{S_o} \tag{3.9}$$

where,

S_o = influent substrate concentrate, mg/L
S_e = effluent/reactor substrate concentration, mg/L
K = complete mix reaction rate coefficient, day^{-1}

This relationship is shown in Figure 3.9. The performance and operating characteristics of the complete mix activated sludge (CMAS) process, when treating several industrial wastewaters, is shown in Table 3.1. The reaction rate coefficients [K in Equation (3.8)] for a variety of wastewaters are summarized in Table 3.1.

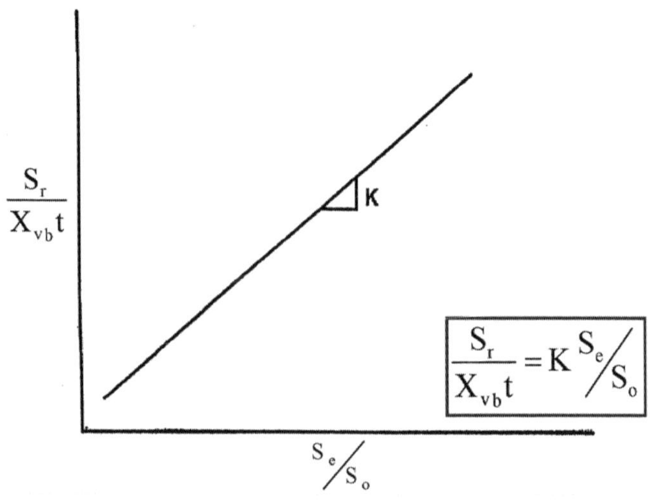

Figure 3.9. Biodegradation coefficient (K) (complete mix system).

In some cases, a two-stage complete mix activated sludge process has been employed for removal of organics from industrial wastewaters. Since the more readily degradable organic components will be removed in the first stage, the rate coefficient, K, will decrease in the second stage. The difference in rate coefficients is shown in Figure 3.10 for a synthetic fiber wastewater. If the only

TABLE 3.1. Reaction Rate Coefficient for Selected Wastewaters [4].

Wastewater Source	K (d^{-1})	Temperature (°C)
Vegetable tannery	1.2	20
Cellulose acetate	2.6	20
Peptone	4.0	22
Organic phosphates	5.0	21
Vinyl acetate monomer	5.3	20
Organic intermediates	5.8	8
Viscose rayon and nylon	20.6	26
	6.7	11
	8.2	19
Domestic Sewage (soluble)	8.0	20
Polyester fiber	14.0	21
Formaldehyde, propanol, methanol	19.0	20
High-nitrogen organics	22.2	22
Potato processing	36.0	20

concern is total organics (i.e., BOD or COD) in the effluent, then there is little advantage to a two-stage process. This is indicated in Figure 3.10 where the reaction rate coefficient data for the single-stage CMAS are also shown (but no linear trace is drawn). However, if specific constituents (e.g., phenol) must be removed to low effluent concentrations, a second stage may be of considerable advantage.

In the second stage, the slower growing "substrate specific" biomass can be concentrated by operating at a higher SRT than would be possible in a single-stage or the first-stage reactor due to MLSS buildup. Furthermore, if the specific substrate removal rate is first-order (or higher), the effluent concentration will be lower in a two- or multiple-stage reactor than in a single-reactor system.

Variability of K-Rates

It is apparent that, as the organic composition of the wastewater changes, the rate coefficient K in Equation (3.3) will also change. This is not a problem for wastewaters such as those from a dairy or food processing plant since their composition remains substantially unchanged, and hence, K will remain nearly constant. Wastewaters generated from plants with multi-products and campaign production, however, will experience a constantly changing wastewater composition and, hence, a highly variable K value. The rate coefficient

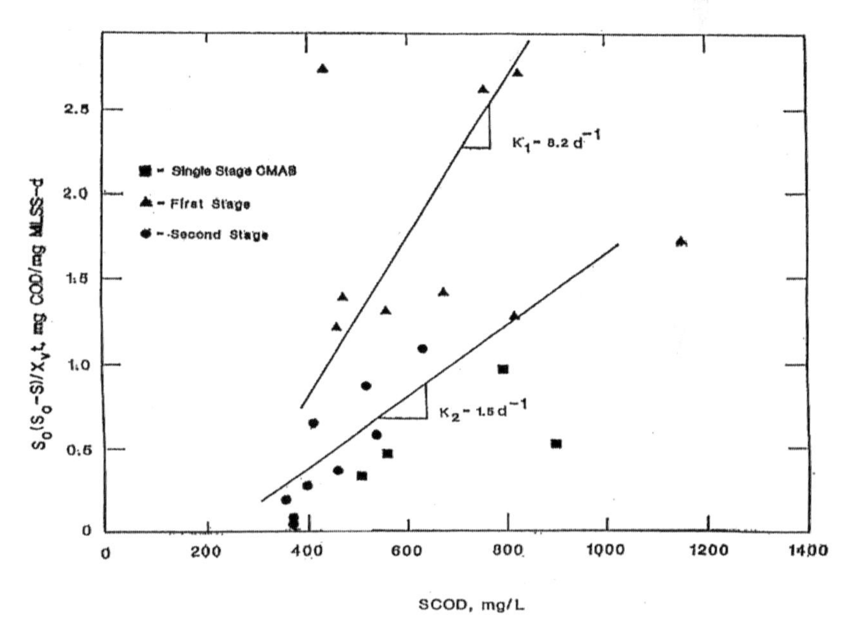

Figure 3.10. Effect of two-stage operation on reaction rate coefficient for a synthetic fibers wastewater [4].

combines the effects of all removal mechanisms: biosorption, biodegradation, and volatilization, unless steps are taken to separate the effects of an individual removal mechanism. Unusually high "apparent" reaction rate coefficients may be observed when volatile organics constitute a large portion of the wastewater. Volatilization of substrate should be considered when calculated K values exceed about 30/day at 20 to 25°C.

Soluble Microbial Products

Soluble microbial products (SMP) are generated in the activated sludge process through the biodegradation of organics and through endogenous respiration of the biomass. The SMP are oxidation by-products that are nondegradable. Pitter and Chudoba [6] have indicated that, depending on cultivation conditions, the nonbiodegradable waste products call amount to 2 to 10 percent of the COD removed. Data for biodegradation of a peptone-glucose mixture and a synthetic fiber wastewater are shown in Figure 3.11. They indicate that approximately 0.20 mg of nondegradable TOC (TOC_{nd}) was produced per mg of influent TOC for the synthetic fiber wastewater. The peptone-glucose containing wastewater produced approximately 0.12 mg TOC_{nd} per mg influent TOC. In both cases, the ratios were constant over the range of influent loading conditions. This indicates that there was a constant metabolic by-product or a portion of the original substrates that were nondegradable.

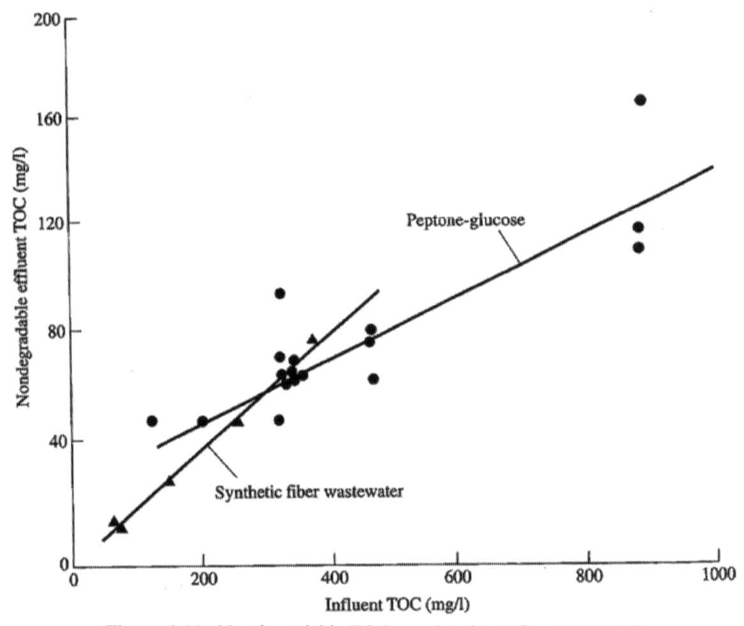

Figure 3.11. Nondegradable TOC as related to influent TOC [4].

Many of the metabolic by-products are of high molecular weight. Approximately 75 percent of the SMP_{nd} from treatment of a wastewater containing only phenol had a molecular weight above 1,000 [7]. It has been further determined that some of these high molecular weight fractions are toxic to some aquatic organisms. There is reason to believe that the high-molecular weight SMPnd strongly adsorb on activated carbon. This characteristic makes granular- (GAC) or powdered-activated carbon (PAC) an excellent candidate process for toxicity reduction when toxicity is caused by SMP_{nd}.

TEMPERATURE

The effect of temperature on the activated sludge process is illustrated in Figure 3.12. The magnitude of the effect is related to the characteristics of the wastewater organics and their physical state (i.e., suspended, colloidal, or soluble). The effect of mixed liquor temperature on the reaction rate coefficient can be expressed as:

$$K_T = K_{20}\theta_K^{(T-20)} \tag{3.10}$$

where,

K_T, K_{20} = reaction rate coefficients at mixed liquor temperatures of T and 20°C, respectively, day^{-1}
T = mixed liquor temperature, °C
θ_K = empirical temperature correction coefficient, dimensionless

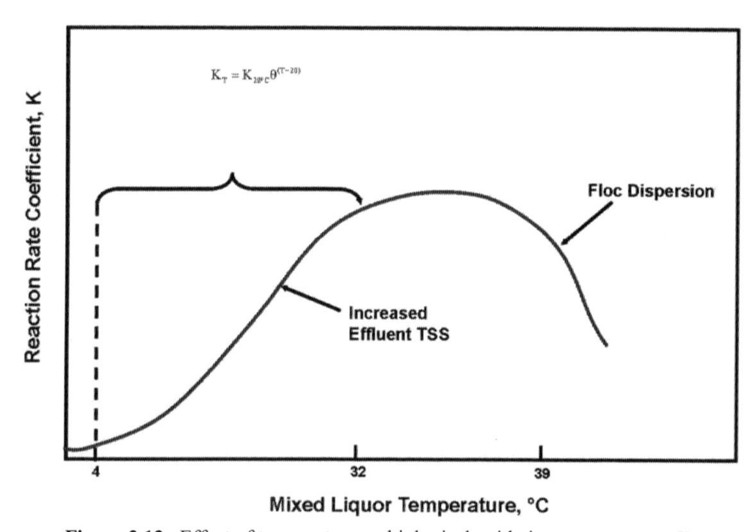

Figure 3.12. Effect of temperature on biological oxidation rate constant, *K*.

Equation (3.10) indicates that the K value increases exponentially with temperature and is applicable over a temperature range of 4 to 31°C. At mixed liquor temperatures between 31 and 39°C, the value of K is approximately constant and then declines at higher temperatures. The decline in the rate coefficient at a temperature above about 40°C is frequently paralleled by a deterioration and dispersion of the biological floc, poor sludge settleability, and high effluent suspended solids and turbidity. For industrial wastewaters containing high soluble substrate concentrations, the value of the temperature correction coefficient (θ_K) has been found to vary from 1.03 to 1.10. Since no consistent correlation has been demonstrated between the value of θ_K and wastewater characteristics, it is necessary to determine θ_K experimentally.

It should be noted that a significant difference in the effect of temperature on the substrate removal rate has been found for municipal wastewater, as compared to industrial wastewaters. This is because a primary substrate removal mechanism for municipal wastewater is the biological entrapment of suspended and colloidal organics—a physical phenomenon that is generally independent of temperature. As a result, the temperature correction coefficient for municipal wastewater treatment has been found to have a value of 1.015 when considering overall BOD removal. The temperature effect on removal of specific organic compounds is frequently much greater than its effect on overall organic removal.

The magnitude of the temperature effect on removal of specific compounds creates operating problems when low effluent values are required on a year-round basis. In many cases, permit values are readily achievable during summer operation but difficult or impossible to achieve during winter operation. The application of steam to raise the mixed liquor temperature is frequently required during cold weather conditions. If the substrate removal rate is inhibited by a toxic agent, the extent of inhibition is frequently increased during cold weather operating conditions. Under these circumstances, powdered-activated carbon can be seasonally added to reduce the bioinhibitory effect.

In addition to the temperature effects on the biological reaction rate, two other temperature-related phenomena must be considered. As the mixed liquor temperature decreases from approximately 25°C towards 5–8°C, there may be an increase in effluent suspended solids. These suspended solids are typically dispersed in nature, and are not removed by conventional clarification processes. For example, a multi-product organic chemicals plant had an average effluent suspended solids value of 42 mg/L during the summer when the average mixed liquor temperature was 28°C. During the winter, the average mixed liquor temperature was 15°C, and the average effluent suspended solids was 104 mg/L. The second temperature-related effect occurs as the mixed liquor temperature increases above about 35°C. At these temperatures, the biological floc characteristics deteriorate is shown in Figure 3.13. At 110°F, however, there were no protozoa, the floc was dispersed, and filamentous forms were

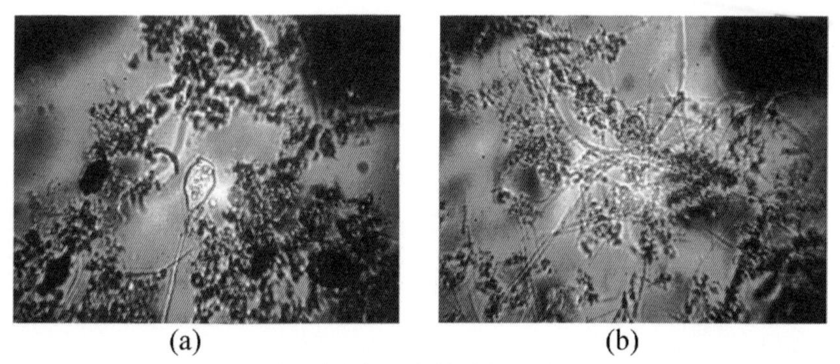

Figure 3.13. Activated sludge at 35°C (96°F) and 43°C (110°F) [4].

present. This phenomenon appears to be related to both the temperature and the characteristics of the wastewater. Floc dispersion and loss of settleability was observed at 41°C for a pulp and paper mill wastewater but occurred at 35°C for an agricultural chemicals wastewater [8]. The increase in effluent suspended solids with mixed liquor temperature for this wastewater is shown in Figure 3.14. At 96°F, the flocs were large, and protozoa were present.

Most activated sludge processes operate in the mesophilic range (4 to ≈39°C). Efforts, however, have been made to operate the activated sludge process in the thermophilic range (45–55°C) when the influent wastewater is already in this temperature range and the BOD is high (2,500–3,000 mg/L). Under these operating conditions, the sludge generated was frequently difficult to separate, the mixed liquor was dispersed, and effluent solids concentrations

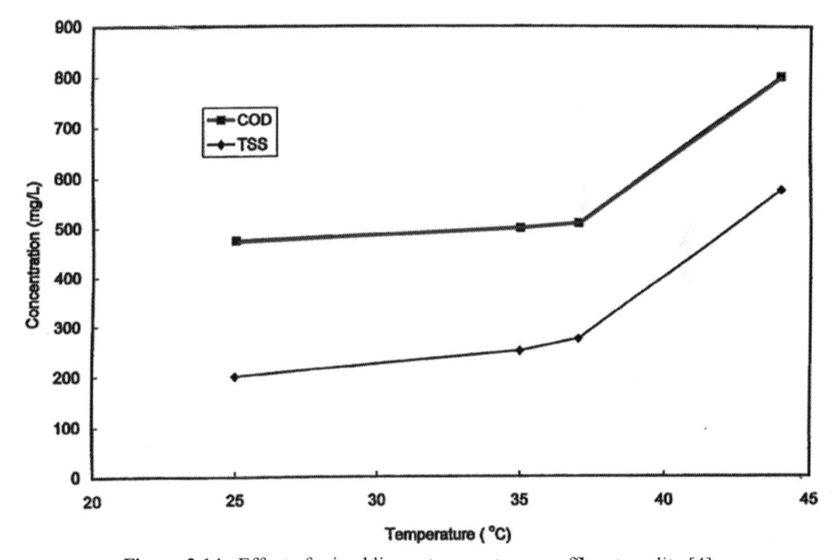

Figure 3.14. Effect of mixed liquor temperature on effluent quality [4].

were high. Thermophilic operating conditions will result in lower sludge production rates and reduced aeration basin tankage requirements. Bench- and/or pilot-scale studies should be conducted to determine sludge settling quality and oxygen transfer characteristics at the elevated temperatures.

SLUDGE PRODUCTION

Using the above relationships, the sludge yield for a wastewater that contains primarily soluble organic substrates can be expressed by Equation (3.11). This relationship is shown in Figure 3.15 for a soluble wastewater from pharmaceutical manufacturing.

$$\Delta X_v + aS_r - bX_d X_v t \tag{3.11}$$

where,

a = biomass yield coefficient, mg VSS/mg BOD (or COD)
S_r = organic removal, mg/L
X_d = biodegradable fraction of biomass
b = endogenous decay coefficient, day^{-1}

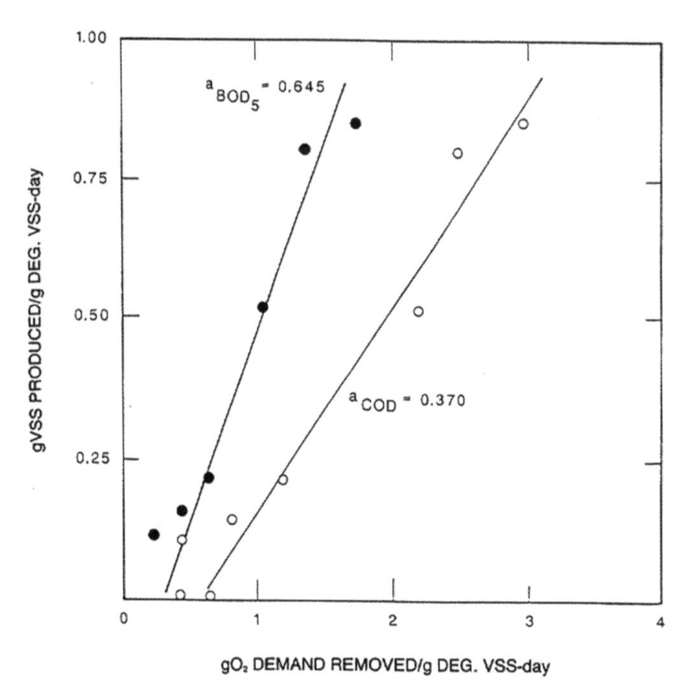

Figure 3.15. Cell synthesis relationship for a soluble pharmaceutical wastewater [4].

The biomass generated in Equation (3.10) is about 80 percent biodegradable. As the SRT is increased, the degradable portion of the biomass will be endogenously oxidized, and the biodegradable fraction of the remaining volatile biomass (designated as X_d) will decrease. The degradable fraction of the volatile biomass can be calculated for a wastewater containing only soluble substrates by Equation (3.12).

$$X_d = \frac{0.8}{1 + 0.2b\theta_c} \tag{3.12}$$

Equation (3.13) can be used to determine X_d for a soluble organic substrate using the kinetic coefficients:

$$X_d = \frac{(aS_r + bX_v t) - [(aS_r + bX_v t)^2 - (4bX_v t)(0.8aS_r)^{0.5}]}{(2bX_v t)} \tag{3.13}$$

If the influent contains VSS, such as in a pulp and paper mill wastewater, Equation (3.14) is modified to include this contribution.

$$\Delta X_v = a[S_r + f_d f_x X_i] - bX_d f_b X_v t + (1 - f_d) f_x X_i + (1 - f_x) X_i \tag{3.14}$$

where,

X_i = influent VSS, mg/L
f_x = fraction of influent VSS that is degradable
f_d = fraction of degradable influent VSS degraded
f_b = fraction of mixed liquor VSS that is biomass

The degradation rate of the influent degradable VSS is a function of the SRT and their specific degradation rate.

$$(1 - f_d) = e^{-K_p \theta_c} \tag{3.15}$$

where,

K_p = degradation rate coefficient of influent VSS, day^{-1}

If it is assumed that 1 mg/L VSS solubilizes to generate 1 mg/L COD, then the fraction of biomass (f_b) in the overall mixed liquor can be determined as follows:

$$f_b = \frac{a[S_r + f_d f_x X_i] - bX_d f_b X_v t}{a[S_r + f_d f_x X_i] - bX_d f_b X_v t + (1 - f_d) f_x X_i + (1 - f_x) X_i} \tag{3.16}$$

Most pulp and fiber in pulp and paper mill wastewaters are essentially nondegradable, and hence, $(1 - f_d)$ is approximately unity. In food processing wastewaters, however, $(1 - f_d)$ may be less than 0.2. If the influent contains high levels of VSS, the value $(1 - f_d)$ must be experimentally determined in order to accurately predict the volatile sludge production rate and true biomass yield. When the wastewater contains influent VSS, Equation (3.13) for X_d must be modified as follows:

$$X_d = \frac{J - [J^2 - (4bf_b X_v t)(0.8aS_r)]^{0.5}}{2bf_b X_v} \tag{3.17}$$

where,

$J = aS_r + bf_b X_v t - f_d f_x X_i$

The impact of influent nonvolatile suspended solids on mixed liquor characteristics and sludge production rate can also be significant. The quantity of inert material (measured as nonvolatile suspended solids) generated is related to the SRT, the hydraulic retention time, the fraction of influent nonvolatile suspended solids that is nondegradable/non solubilized, and the formation of nonbiomass particulates in the activated sludge process. This relationship is expressed as follows:

$$\Delta \text{NVSS} = a^* S_r f_{\text{ibnd}} + f_{oi} X_{oi} + \text{Nonbiomass particulate formation} \tag{3.18}$$

where,

ΔNVSS = inert suspended solids produced, mg/L
X_{oi} = influent inert solids, mg/L
f_{ibnd} = fraction of inert biomass
f_{oi} = fraction of influent inert solids not degraded or solubilized
a^* = biomass produced per unit of substrate removed, mg TSS/mg BOD (or COD)

Inert material may also be produced in the activated sludge system through precipitation reactions of the wastewater. This later accumulation term is indicated but not characterized in Equation 3.18 since it is difficult to quantify unless the influent suspended solids concentration is negligible and the inert accumulation is significant.

Total sludge production can be calculated by summing Equations (3.11) and (3.18). The impact of the total suspended solids production on the operating aeration basin MLSS concentration is a direct function of SRT. As SRT increases, the aeration basin MLSS will increase. Figure 3.16 shows the observed sludge yield as a function of sludge age. Consideration must be given in secondary clarifier solids loading rate design to accommodate the inert and volatile suspended solids generation while maintaining the required SRT for

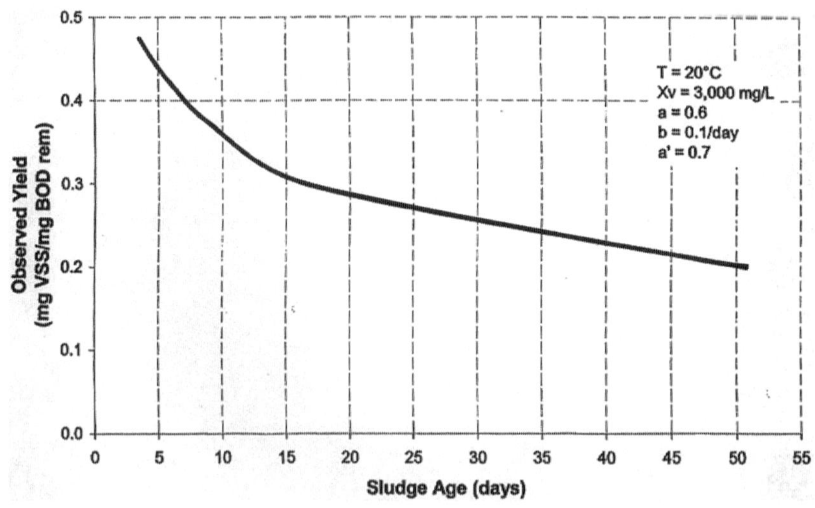

Figure 3.16. Observed sludge yield versus sludge age.

substrate removal. Example 3.1 presents a design example which illustrates how to calculate the sludge production.

Example 3.1—Activated Sludge Design Example

Determine the operating F/M, MLVSS, fb and sludge production for an activated sludge process under the following operating conditions:

$a = 0.45$
$b = 0.1$ at 20°C
$\theta_c = 10$ days
$X_i = 200$ mg/L
$f_x = 0$
$S_o = 1{,}000$ mg/L
$S_e = 20$ mg/L
$t = 0.9$ days

1. The F/M is calculated as follows:

$$\frac{1}{\theta_c} = a(F/M) - bX_d$$

$$X_d = \frac{0.8}{1 + 0.2b\theta_c}$$

$$X_d = \frac{0.8}{1 + 0.2(0.1 \cdot 10)} = 0.67$$

$$\frac{1}{10} = 0.45(F/M) - (0.1 \cdot 0.67)$$
$$F/M = 0.37/\text{day}$$

2. The MLVSS (X_v) is composed of biomass VSS and influent VSS that are not biodegradable and is calculated as

$$F/M = S_o/(f_b X_v t)$$
$$f_b X_v = 1{,}000/(0.37)(0.9)$$
$$f_b X_v = 3{,}000 \text{ mg/L}$$
$$X_v = [(aS_t - bX_d f_b X_v t)/(1 - f_x)X_i]\frac{\theta_c}{t}$$
$$= [0.45(1{,}000 - 20) - (0.1 \cdot 0.67 \cdot 0.9 \cdot 3{,}000) + (1.0 - 0)200]\frac{10}{0.9}$$
$$X_v = 5{,}112 \text{ mg/L}$$

3. The biomass fraction of X_v is

$$f_b = \frac{3{,}000}{5{,}112} = 0.59$$

4. The waste sludge production rate per pass through the aeration basin is

$$\Delta X_v = aS_r - bX_d f_b X_v t + \Delta X_i$$
$$= (0.45 \cdot 980) - (0.1 \cdot 0.67 \cdot 3{,}000 \cdot 0.9) + (1.0 - 0)200]$$
$$= 441 - 181 + 200$$
$$\Delta X_v = 460 \text{ mg/L}$$

5. Check that $\theta_c = 10$ days

$$\theta_c = \frac{5{,}112 \cdot 0.9}{460} = 10 \text{ days}$$

Example 3.2—Activated Sludge Design Example

Figure 3.17 shows a mass balance diagram for this activated sludge design example. This design example illustrates the basic design calculations required to develop the designs for an activated sludge treatment plant. A Biowin model could also be setup to develop the same design and facilitate sensitivity analy-

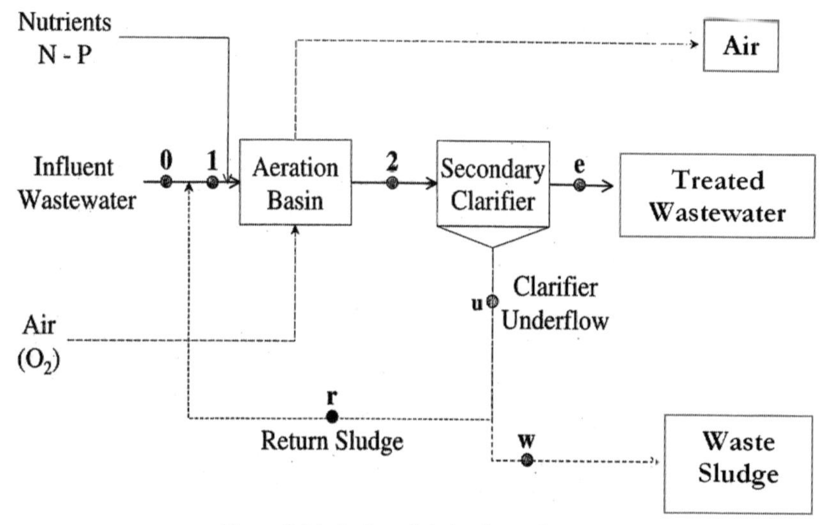

Figure 3.17. Activated sludge flow schematic.

sis to compare the designs over a range of process parameters such as temperature and sludge age.

Influent COD

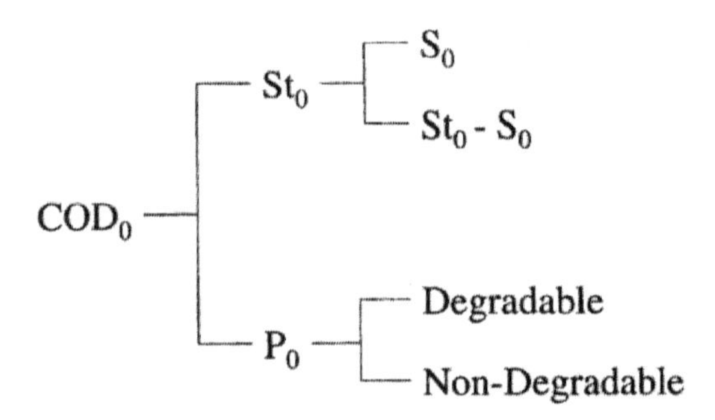

Design Information

$$S_{t0} = 500 \text{ mg/L} \qquad S_0 = 0.9 \cdot S_{t0} \qquad S_0 = 450 \text{ mg/L}$$

$$S_e = 40 \text{ mg/L}$$

$$a = 0.45 \qquad a' = 0.37$$

$$K_{20} = 7.5/\text{day} \qquad b_{20} = 0.1/\text{day}$$

$$T = 20°C$$

$$X_v = 2500 \text{ mg/L}$$

Kinetic Parameters

$$b = b_{20} \cdot 1.04^{(T-20)} = \frac{0.1}{day}$$

$$K = K_{20} \cdot 1.09^{(T-20)} = \frac{7.5}{day}$$

1. Degradable and Active Fraction of Biological MLVSS

$$X_d = \frac{0.8}{(1+0.2 \cdot b \cdot \theta_C)} = \frac{0.8}{(1+1.2 \cdot 0.1 \cdot 5)} = 0.727$$

$$f_a = \frac{(X_d)}{0.8} = 0.909$$

2. Sludge Age (θ_C)

$$\theta_C = \frac{S_0}{a \cdot K \cdot S_e \cdot b \cdot S_0} = \frac{450}{0.45 \cdot 7.5 \cdot 40 - 0.1 \cdot 450} = 5 \text{ days}$$

3. Detention Time (t)

$$t = \frac{(S_0 - S_e) \cdot S_0}{K \cdot f_a \cdot X_V \cdot S_e} = \frac{(450-40) \cdot 450}{7.5 \cdot 0.91 \cdot 2500 \cdot 40} = 0.271 \text{ day}$$

4. Oxygen Requirements (R_{O_2})

$$R_{O_2} = a' \cdot (S_0 - S_e) + 1.4 \cdot b \cdot \chi_V \cdot X_v \cdot t$$

$$R_{O_2} = 0.37 \cdot (450-40) + 1.4 \cdot 0.1 \cdot 0.73 \cdot 2500 \cdot 0.271 = 221 \text{ mg/L}$$

5. Waste Activated Sludge (ΔX_V)

$$\Delta X_v = a \cdot (S_0 - S_e) - b \cdot \chi_d \cdot X_V \cdot t$$

$$\Delta X_v = 0.45 \cdot (450-40) - 0.1 \cdot 0.73 \cdot 2500 \cdot 0.271 = 135 \text{ mg/L}$$

$$\Delta XH = \frac{\Delta XVH}{0.8} = 169 \text{ mg/L}$$

6. Nutrients Requirements (N, P)

$$N = [0.123 \cdot \chi_d + 0.07 \cdot (0.8 - \chi_d)] \cdot \frac{\Delta X_V}{0.8}$$

$$N = [0.123 \cdot 0.73 + 0.07 \cdot (0.8 - 0.73)] \cdot \frac{135}{0.8} = 16 \text{ mg/L}$$

$$P = [0.026 \cdot \chi_d + 0.01 \cdot (0.8 - \chi_d)] \cdot \frac{\Delta X_V}{0.8}$$

$$P = [0.026 \cdot 0.73 + 0.01 \cdot (0.8 - 0.73)] \cdot \frac{135}{0.8} = 3.3 \text{ mg/L}$$

7. Mixed Liquor Solids (X)

$$X = \frac{X_v}{0.8} = \frac{2500}{0.8} = 3125 \text{ mg/L}$$

OXYGEN UTILIZATION

The biological oxygen requirements can be computed by Equation (3.19):

$$O_2 = a'S_r = (1.4b)X_d f_b X_v t \tag{3.19}$$

where,

O_2 = oxygen requirement, mg/L
$f_b X_v$ = biomass under aeration, mg VSS/L

The "lumped" coefficient, $1.4b$, is frequently referred to as b'. Equation (3.19) is used to determine a' and b' and the system oxygen requirements. This relationship is shown in Figure 3.18 for a food processing wastewater where

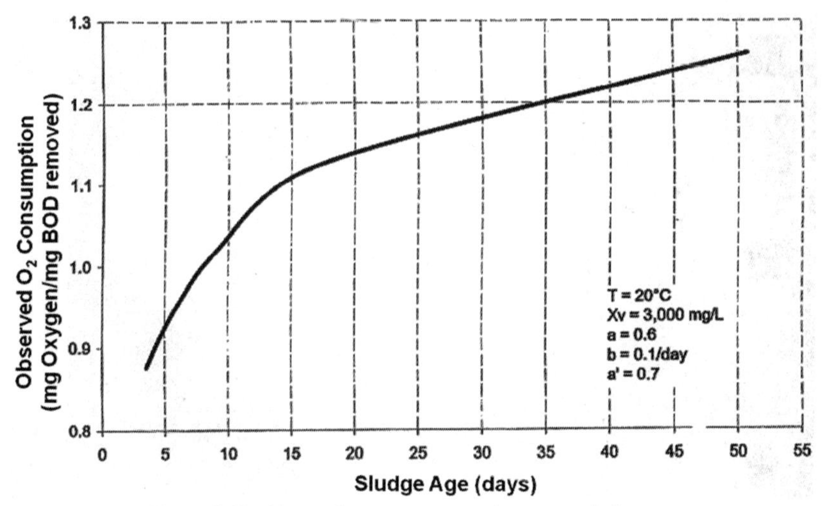

Figure 3.18. Observed oxygen consumption versus sludge age.

$f_b \approx 1$. The oxygen utilization coefficients, a' and b', and the sludge yield coefficient, a, can be determined from wastewater treatment plant operating data using the above relationships. Figure 3.18 shows the observed oxygen consumption as a function of sludge age.

NUTRIENT REQUIREMENTS

The biomass requires nitrogen and phosphorus in order to effect synthesis, metabolism, and removal of organics in the treatment process. In addition to these "bulk nutrients;" trace levels of other nutrients are required to assure good floc formation. Trace nutrient requirements are shown in Table 3.2. These are usually present in sufficient quantities in the carrier water of most process waste streams, except in cases of high-strength wastewaters and/or where deionized water is used in production and constitutes the carrier for the process wastewater. Addition of small amounts of iron (and sometimes other nutrients) will usually be required under these conditions.

When insufficient nitrogen is present, the amount of cellular material synthesized per unit of organic matter removed increases due to an accumulation of polysaccharide. At some point, nitrogen-limiting conditions become severe and restrict the rate of BOD removal. Nutrient-limiting conditions will also stimulate filamentous growth.

Nitrogen is available to the biomass in the preferred form as ammonium (NH_4^+) or as nitrate (NO_3^-). Organic nitrogen, present in the wastewater as protein or amino acids, must first be biologically hydrolyzed to release ammonium in order to be available to the biomass. Therefore, in wastewaters containing organic nitrogen as the primary nitrogen source, experiments must be

TABLE 3.2. Trace Nutrient Requirements for Activated Sludge [4].

Micronutrient	Requirement (mg/mg BOD)
Mn	10×10^{-5}
Cu	15×10^{-5}
Zn	16×10^{-5}
Mo	43×10^{-5}
Se	14×10^{-10}
Mg	30×10^{-4}
Co	13×10^{-5}
Ca	62×10^{-4}
A	5×10^{-5}
K	45×10^{-4}
Fe	12×10^{-3}

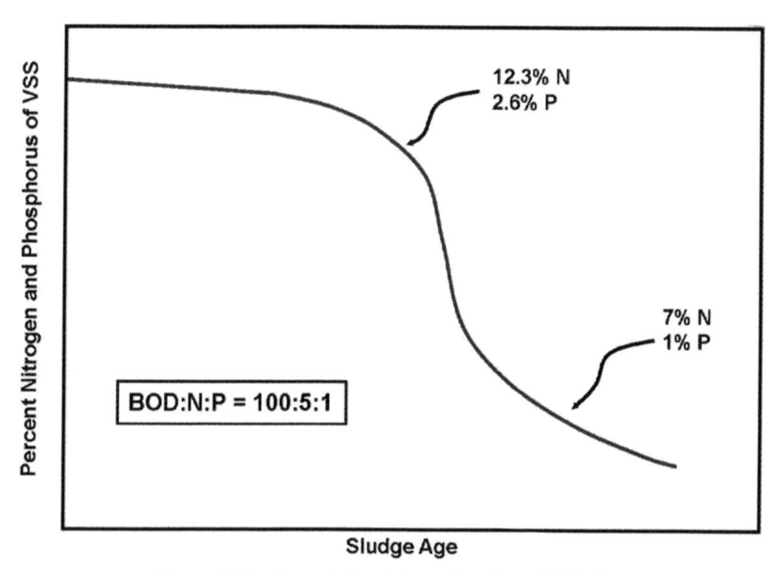

Figure 3.19. Percent *N* and *P* as a function of SRT [4].

conducted to determine the availability of the organic nitrogen for biomass utilization since some aromatic amines and tertiary aliphatic amino compounds are not readily hydrolyzed.

Phosphorus must be in the form of soluble orthophosphate (o-PO_4) in order to be assimilated by the biomass. Therefore, complex inorganic and organically bound phosphorus must first be biohydrolyzed to orthophosphate in order to be available for biosynthesis.

The "rule of thumb" to assure adequate nitrogen and phosphorus for BOD removal is to provide a maximum nutrient mass ratio of 100:5:1 (BOD:N:P). A higher ratio (e.g., 150:5:1) will reduce the rate of BOD removal and promote filamentous growth. In a continuous flow process, however, the actual nitrogen and phosphorus requirement will depend on the net biomass synthesis (i.e., nitrogen assimilation due to growth and nitrogen release through endogenous respiration). This balance will be related to the nitrogen and phosphorus content of the wasted sludge as determined by the wastewater characteristics and the SRT. The nitrogen content of the biological sludge, as generated in the process under nitrogen-rich conditions, is approximately 12.3 percent (by weight) based on the VSS. However, the nitrogen content of the sludge declines in the endogenous phase and when nitrogen is limiting growth. The nitrogen contest of the nondegradable cellular mass has been shown to average 7 percent. These relationships are shown in Figure 3.19. In like manner, the phosphorus content of sludge at nongrowth limiting conditions has been found to average 2.6 percent, with the nondegradable cellular residue having a phosphorus content of approximately 1 percent.

The nitrogen and phosphorus requirements can be calculated by Equations (3.20) and (3.21), respectively, considering the nitrogen and phosphorus content of the biomass wasted from the process.

$$N = 0.123\frac{X_d}{0.8}\Delta X_v + 0.07\frac{(0.8 - X_d)}{0.8}\Delta X_v \tag{3.20}$$

$$P = 0.026\frac{X_d}{0.8}\Delta X_v + 0.01\frac{(0.8 - X_d)}{0.8}\Delta X_v \tag{3.21}$$

ACCLIMATION

When treating industrial wastewaters, particularly for removal of specific organics, it is necessary to acclimate the biomass to the wastewater. The source of the seed biomass (i.e., municipal or an industrial sludge treating a similar wastewater), the operating temperature, and the sludge age will determine the time required for acclimation. Typical acclimation periods are from several days to five to six weeks, or acclimation may not occur at all. Acclimation results obtained by Tabak *et al.* [9] for several organic compounds, when starting with a municipal biomass as seed, are shown in Figure 3.20.

If the wastewater is readily degradable and susceptible to filamentous bulk-

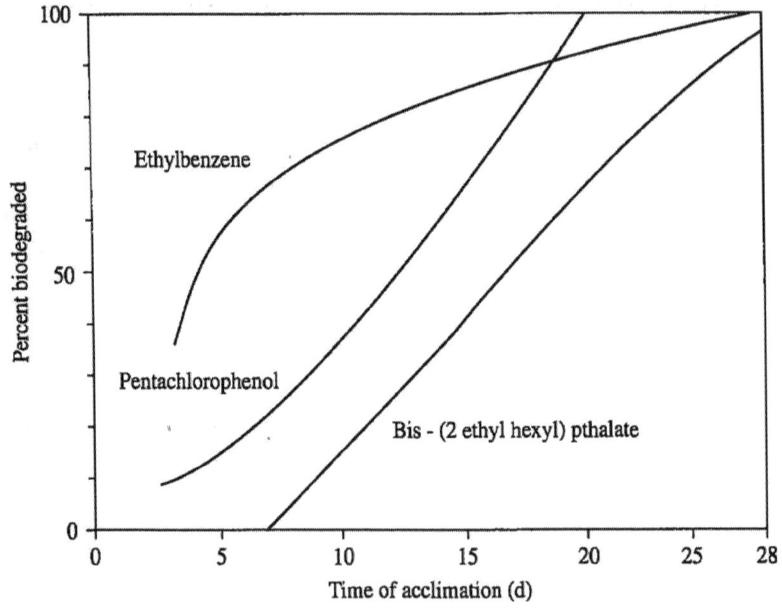

Figure 3.20. Acclimation of activated sludge to specific organics [4].

ing, acclimation of the biomass is best achieved on a batch (fill and draw) operating basis. After acclimation is established, the operation can be converted to a continuous flow basis using either a plug-flow mixing regime or a biological selector to control bulking. A wastewater containing biorefractory organics can be acclimated in a continuous flow system. If the wastewater is bioinhibitory, acclimation must start at concentrations well below the inhibition threshold concentration. This concentration can be defined using the fed-batch reactor (FBR) or other similar test procedures as discussed in Chapter 7. As acclimation proceeds and biodegradation commences, the concentration of the target wastewater can be gradually increased as long as the concentration in the reactor does not exceed the inhibition threshold. Acclimation of the biomass to a specific wastewater composition is assumed to be complete when the specific oxygen uptake rate or the residual organic concentration reaches a steady-state condition.

BIOINHIBITION

Many organics will exhibit a threshold concentration at which they inhibit the heterotropic and/or nitrifying organisms in the activated sludge process. Inhibition has been defined by the Haldane equation (or its modifications) using Monod kinetics:

$$\mu = \frac{\mu_m S X_v}{S + K_s + S^2 / K_1} - b X_d X_v \tag{3.22}$$

in which K_1 is the Haldane inhibition coefficient.

An example of bioinhibition by a plastics additives [4] wastewater showed that the concentration of influent COD (and inhibitory agent) increased, the SOUR decreased, resulting in higher effluent SBOD concentrations. In this case, the inhibition was removed by pretreating the wastewater with hydrogen peroxide (H_2O_2), thereby effecting detoxification and enhanced biodegradability. Adding powdered activated carbon to the mixed liquor can be used to absorb the toxicant (see PACT Technology in Chapter 4).

Volskay and Grady [10] and Watkin [7] have shown that inhibition can be competitive (the inhibitor affects the base substrate utilization), noncompetitive (the inhibitor rate is influenced), or mixed, in which both rates are influenced. The effects of substrate and inhibitor concentrations on the respiration rate of a microbial culture expressed as a fraction of the rate in the absence of the inhibitor have been shown by Volskay and Grady [10].

While the relationships described above define the mechanism of inhibition, they are of limited use in evaluating industrial wastewaters. In most cases, the inhibitor itself is not defined, variable sludge and substrate composition will influence inhibition, and interactions will frequently exist between inhibitors. The inhibition constant, K_I, is highly dependent on the specific enzyme system involved, which, in turn, is dependent on the history and population dynamics of the sludge. In some cases, the inhibition constant may be dependent on the particular metabolic pathways that are present in any given microbial population. For example, Watkin and Eckenfelder [8] showed a variation in K_I of 6.5 to 40.4 for different sludges and operating conditions treating 2,4-dichlorophenol and glucose. Volskay and Grady [10] showed a variation of 2.6 to 25 mg/L in the concentration of pentachlorophenol, which would cause 50 percent inhibition of oxygen utilization rates.

It is apparent, therefore, that each wastewater must be independently evaluated for its bioinhibition effects. Several protocols have been developed for this purpose and are presented elsewhere [3]. These include the Fed-Batch Reactor of Philbrook and Grady [11] and Watkin and Eckenfelder [8], the OECD Method 209 of Volskay and Grady [10], and the Glucose Inhibition Test of Larson and Schaeffer [12]. The authors have been using the Fed-Batch Reactor and other batch tests protocols over the last ten years and these are discussed in Chapter 7.

FINAL CLARIFICATION

Final Clarification is key to achieving separation of the treated wastewater from the biomass and performance of the activated sludge process. The evolution of membranes in a membrane bioreactor to replace final clarification is discussed in Chapter 4. The activated sludge process should be designed and operated at an *F/M* and sludge age to both optimize removal of BOD and other constituents and to achieve optimizing flocculation of the biomass in order to achieve good settleability of the sludge in the final clarifier (see previous Figure 3.3). The final clarification process is also key to achieve thickening of the biomass recirculated back to the aeration basin. A portion of the thickened biomass is wasted daily or at some frequency to achieve the desired sludge age. Final clarification design is discussed in other books [1,2] and references [13]. Final clarification, along with effluent suspended solids, is key to achieve the effluent permit limits for TSS as well as other constituents which may be adsorbed to the suspended solids.

CONTROL OF ACTIVATED SLUDGE QUALITY

Activated sludge characteristics are shown in Figure 3.21. The non-bulking sludge floc in the middle with filament backbone is the optimum quality. Filamentous bulking at the top shows filaments extending beyond the floc.

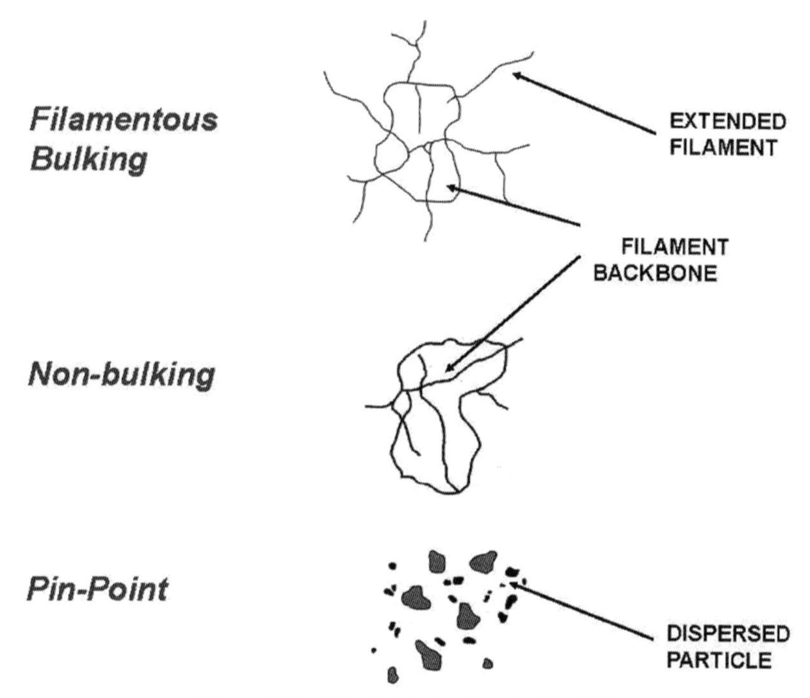

Figure 3.21. Activated sludge characteristics [1].

Pin-point floc or dispersed floc at the bottom can occur at low *F/M* and creating high effluent TSS. When treating industrial and municipal wastewaters, biomass bulking can be caused by the following:

- insufficient mixed liquor dissolved oxygen
- insufficient nutrients
- low *F/M* in the case of readily degradable wastewaters

A deficiency in substrates such as the macro- or micronutrient concentration, residual soluble BOD, and/or dissolved oxygen concentration in the biological floc can promote filamentous growth and sludge bulking. To illustrate these effects, consider the transfer of dissolved oxygen to the hypothetical biological floc particles, as illustrated in Figure 3.22. Oxygen must diffuse from the bulk liquid through the floc in order to be available to the organisms within the interior of the floc particle. As it diffuses, it is consumed by the organisms within the floc. If there is an adequate residual of dissolved oxygen (and nutrients and organics), the rate of growth of the floc-forming organisms will exceed that of the filaments, and a flocculant well-settling sludge will result. If there is a deficiency in any of these substrates, however, the filaments, having a high surface area to volume ratio, will have a "feeding" advantage over the floc formers and will proliferate due to their higher growth rate under the adverse conditions.

Considering Case 1 in Figure 3.22 at an *F/M* of 0.1/day, the oxygen utilization rate is low, and even with a bulk liquid dissolved oxygen concentration of 1.0 mg/L, oxygen will fully penetrate the floc. Under these conditions, the floc formers will outgrow the filaments. In Case 2 the *F/M* is increased to 0.4/day causing a corresponding increase in oxygen uptake rate. If the bulk mixed liquor dissolved oxygen is maintained at 1.0 mg/L, the available oxygen will be rapidly consumed at the periphery of the floc, thus depriving a large interior portion of the floc particle of oxygen. Since the filaments have a competitive growth advantage at low dissolved oxygen levels, they will be favored and will outgrow the floc formers. In a similar manner, insufficient nitrogen and phosphorus concentrations will result in a nutrient deficiency and filamentous bulking.

In a similar manner, the bulk mixed liquor soluble BOD concentration must be sufficient to provide a driving force to penetrate the biological floc. In a complete mix basin, the concentration of soluble BOD in the mixed liquor is essentially equal to the effluent concentration and is, therefore, low (< 10 mg/L) for readily degradable wastewaters. As a result, substrate penetration of the floc is not achieved, and filaments dominate the interior floc population. In order to shift the population in favor of the floc formers, sufficient driving force must be developed to penetrate the floc and favor their growth. This can be achieved by a batch or plug-flow operating configuration in which a high substrate gradient (driving force) exists. Maximum growth of floc formers occurs at the influent end of the plug-flow basin or in the initial period of each feed cycle of a batch-activated sludge process.

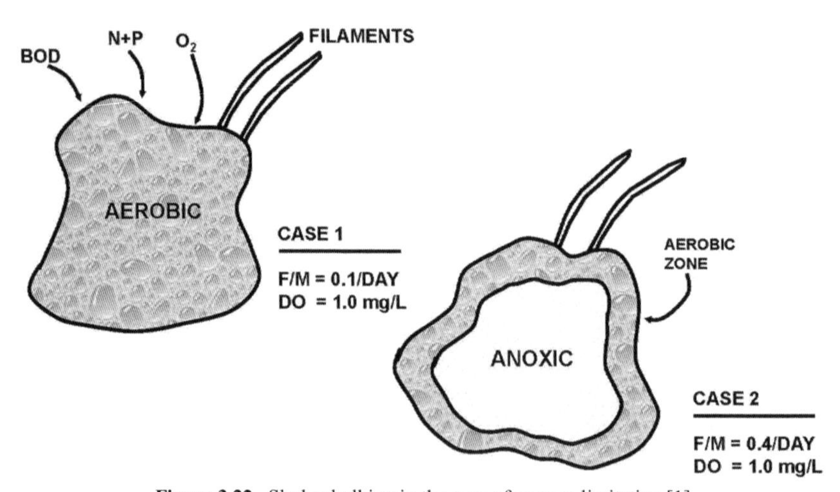

Figure 3.22. Sludge bulking in the case of oxygen limitation [1].

Biological Selectors

A biological selector (see previous Figure 3.4) may be used for filament control instead of a plug-flow or batch treatment process. In the selector, a significant portion of the soluble substrate removal occurs by biosorption. Under these conditions, the substrate gradient is high and promotes the growth of floc formers over the filaments since they have a high "sorption" capacity, whereas the filamentous organisms do not. When the wastewater is discharged from the selector to the downstream CMAS, the soluble substrate concentration is relatively low and is available for utilization by both the floc formers and the filaments. The filaments do not predominate in the mixed liquor, however, since the principal mass of substrate removed in the selector has been initially directed to storage and subsequent growth of floc-forming biomass.

A study of grapefruit processing wastewater [4] compared Reactor Nos. 1 and 2 using an aerobic selector followed by a completely mixed aeration basin versus a plug-low regime aeration basin in Reactor No. 3. In all three cases, the SVI was below 100 mL/g, with effluent BOD concentrations ranging from 4 to 18 mg/L. The plug-flow reactor, however, produced a sludge that was more readily dewatered and had superior thickening properties. A parallel complete mix-activated sludge process (without selector) that was operated at the same organic loading rate as these systems produced severe bulking problems and was shut down since it was inoperable.

Filamentous bulking has also been controlled using an anoxic selector prior to a complete mix-activated sludge process. Since most of the filamentous organisms cannot grow under anoxic conditions, uptake and assimilation of the organics is restricted to the floc-forming organisms. This is illustrated by the results of a parallel study [4] of anoxic and aerobic selectors for treatment of a kraft pulp and paper mill wastewater. The two parallel anoxic selectors had 0.75-hr and 2.5-hr retention times, while the aerobic selector had a 30-min retention time. The effluents from both selectors were treated by oxygen-activated sludge systems. Microscopic analysis of the sludges showed the absence of filaments in the mixed liquor with the anoxic selector (2.5 hr) compared to that with the aerobic selector.

Several methods have been proposed for process design of the aerobic biological selector [1]. Each of these is based on the selector *F/M* or the floc loading relationship for the wastewater-sludge mixture, as defined by previous Equation (3.7). The design objective is to provide sufficient biomass-wastewater contact time to remove a significant portion of the influent degradable substrate. If 60 to 75 percent of the influent degradable substrate is sorbed in the selector, then the subsequent metabolism and growth of the floc formers is usually adequate to establish a well-settling sludge. If less degradable substrate is removed due to an excessive floc load, then higher concentrations "leak" into the activated sludge reactor and support filamentous growth. If the selector floc load is too low, however, filaments may be completely eliminated from the mixed liquor, which will cause low SVI values, but high effluent TSS and turbidity levels. The results of multiple batch floc load tests on a readily degradable pulp and paper mill wastewater are shown in Figure 3.23. These data were used to select a floc loading of 100 to 150 mg COD/g VSS for operation of the aerobic selector of a bench-scale selector-CMAS system.

"Selecting For or Against Organisms" was a topic presented at the June 2013 WEF Forum on "Activated Sludge on Its 100th Birthday: Challenge and Opportunities." The future of activated sludge and the next generation of selectors was a discussion topic. Selecting organisms for short cut nitrogen removal to nitrite-nitrogen rather than nitrate-nitrogen and for phosphorus removal in municipal treatment plants is a new trend being evaluated on projects. Several technologies are presented in Chapter 4.

Chlorination of Sludge

In some cases, increased influent loads result in filamentous bulking that cannot be controlled by the use of a selector or plug-flow regime. It has been found that chlorination of the return sludge will result in a selective kill of the

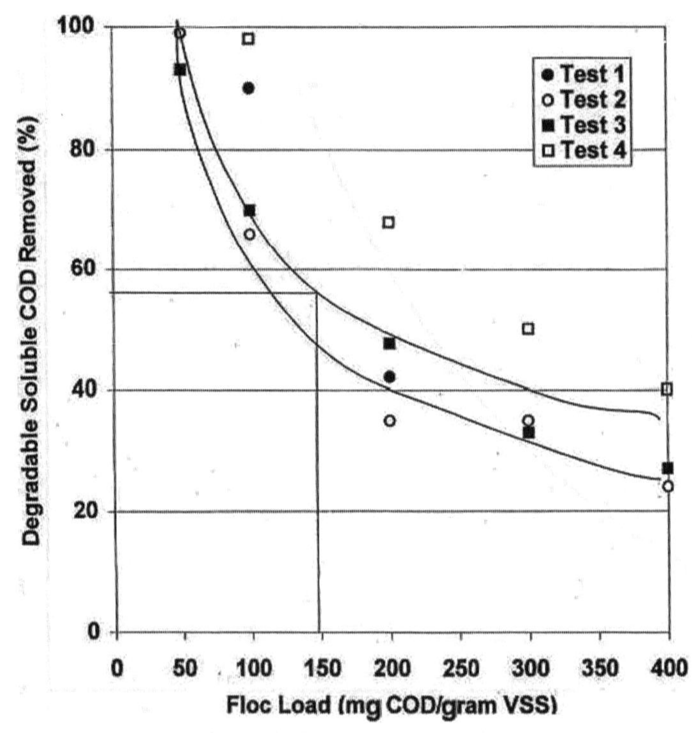

Figure 3.23. Floc results for a recycle paper mill wastewater [4].

filaments. At a controlled dosage of chlorine, the filaments, having a high surface area and exposure to the bulk mixed liquor, will be destroyed (along with those organisms on the periphery of the floc). Filamentous bulking can typically be controlled with chlorine dosages of 10 to 12 lb Cl_2/1000 MLSS-day. It should be noted that if nitrification is required, the chlorine dosage should be restricted to 4.5 lb Cl_2/1000 MLSS-day [4] since the nitrifiers are more sensitive and tend to concentrate on the periphery of the floc where the dissolved oxygen levels are higher.

When treating low strength industrial wastewaters that require a short aeration basin detention time (< 6 to 8 hr), chlorine should be applied to the return sludge. For high-strength industrial wastewaters, however, that require a long hydraulic retention time (> 12 to 16 hr), the chlorine must be directly applied to the mixed liquor. In some cases, hydrogen peroxide has been successfully applied for sludge bulking control. It is probable, however, that this relates to low dissolved oxygen filaments, which are suppressed by the oxygen released by the hydrogen peroxide.

Final clarifier performance is related to the surface overflow rate and solids flux on the clarifier, the return sludge concentration, and the SVI of the sludge. These variables can be correlated using clarifier operational data and the method of Daigger and Roper [13]. It has been found, however, that the sludge-settling properties from industrial wastewater treatment are frequently different from those of domestic sludges.

EFFLUENT SUSPENDED SOLIDS CONTROL

Carryover of suspended solids in the secondary clarifier effluent can be due to several causes including the following:

- floc shear due to high aeration basin power levels
- poor clarifier hydraulics
- high wastewater TDS concentration
- low or high mixed liquor temperature
- rapid change in mixed liquor temperature
- low mixed liquor surface tension
- low *F/M* or high *F/M* (see Figure 3.3)
- denitrification in clarifier and nitrogen gas bubbles

High mixed liquor turbulence levels created by turbine type or mechanical surface aerators can cause floc breakup that results in high-effluent suspended solids. This problem can frequently be solved by reducing the aeration basin power level and/or by installing a flocculation zone between the aeration basin and the final clarifier. High-effluent suspended solids levels will frequently result from a poor clarifier hydraulic design that causes density currents and/or short-circuiting. These conditions result in an upwelling of floc solids at the clarifier peripheral weir. A shallow clarifier with solids carryover which exceeds its discharge limit can be minimized by installing a Stamford baffle, which redirects the upflow of solids away from the effluent weir.

High wastewater TDS levels due to inorganic salts will frequently result in floc dispersion and an increase in effluent suspended solids. Results of treatability studies on an agricultural chemicals wastewater, summarized in Table 3.3, demonstrate that wastewaters at 1 to 2 percent TDS had negligible impact on activated sludge organics removal performance. However, at TDS concentrations of 2 percent, a significant deterioration in effluent TSS quality was observed. Since these solids do not settle, it is necessary to add chemical coagulants to effect their separation. The coagulants can be added between the aeration basin and final clarifier or directly to the final clarifier if adequate mixing, flocculation, and contact time are provided in the clarifier flocculation zone. It is important that turbulence be controlled in the flocculation chamber to avoid excessive floc shearing before entering the clarification zone. Coagulants for bioflocculation include alum, iron salts, and cationic polymer. When treating large volumes of wastewater, cationic polymer is usually more cost-effective when alum or iron dosages exceed 50 mg/L due to lower additional sludge volumes.

The effect of seasonal variation in mixed liquor temperature on the polymer dose required to produce an effluent suspended solids concentration of 40 mg/L is shown in Figure 3.24 for a multi-product organic chemicals wastewater. The decreasing aeration basin temperature resulted in increased polymer dosage to satisfy the required TSS discharge limit. A rapid change (< 24 hr) in mixed liquor temperature will also result in floc dispersion and an increase in effluent suspended solids. This is a temporary effect, however, and mixed liquor settling characteristics will recover when the temperature is stabilized.

Activated sludge may also exhibit poor flocculating characteristics at both

TABLE 3.3. Effects of Total Dissolved Solids on Activated Sludge Treatment of Agricultural Chemicals Wastewater [4].[(a)]

Unit No.	Effluent TDS (mg/L)	Sludge Settleability: Flux Rate (lb/day-sq ft)	Sludge Settleability: SVI (mL/g)	Organics Removal: BOD (%)	Organics Removal: TOC (%)	Settled TSS[(b)] (mg/L)
1	10,600	48	61	94	55	32
2	13,200	51	49	96	53	34
3	15,600	51	47	96	57	38
4	20,200	55	46	93	53	101

[(a)]Units operated at 25°C and $F/M = 0.2$/day.
[(b)]Following 30-min settling period.

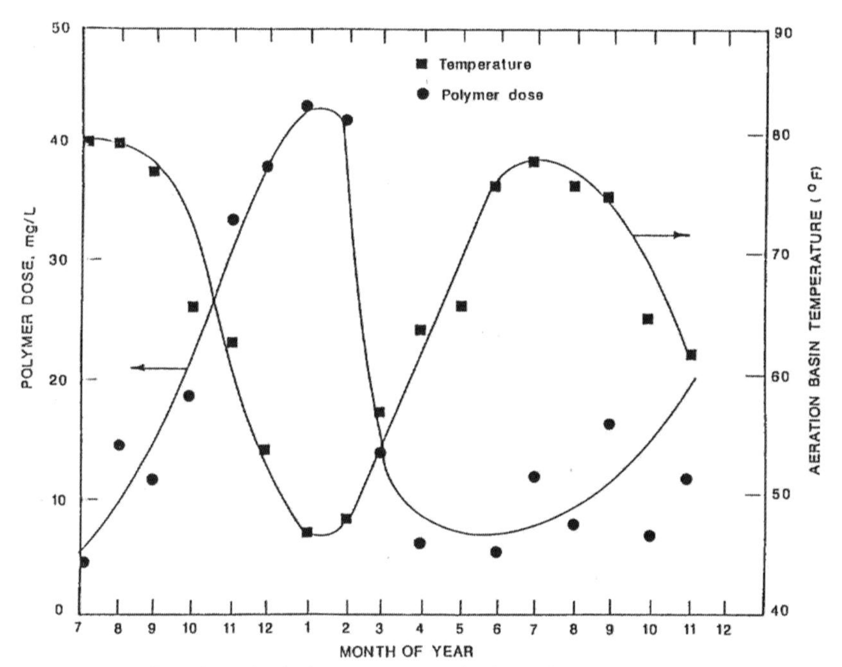

Figure 3.24. Effect of aeration basin temperature and polymer dose required to achieve 40 mg/L suspended solids [4].

very low and very high *F/M* conditions. This effect is demonstrated by increased effluent TSS concentrations, as shown in Figure 3.25 for an organic chemicals wastewater. At *F/M* values between 0.08/day and 0.36/day, the biomass was well flocculated and the effluent TSS was low, as indicated by the probability plot. At low *F/M* (0.04/day), however, bioflocculation deteriorated due to excessive endogenous respiration, and the effluent TSS increased. Similarly, at high *F/M* (0.70/day), bioflocculation and effluent TSS quality were poor.

Dispersed suspended solids also increase with a decrease in surface tension and the presence of chemical dispersants. At one de-inking mill, the effluent suspended solids value was directly related to the surfactant usage in the mill. Similarly, activated sludge floc formation was poor when chemical dispersants were used in a polymer manufacturing process to prevent premature flocculation of the polymer material. The ratio of monovalent to divalent cation in the wastewater can also cause sludge destabilization and an increase of turbidity in the effluent. Magnesium sulfate (divalent cation) addition can be added to decrease the ratio below 2 to improve sludge settleability and reduce effluent TSS.

Technologies for effluent suspended solids control and retaining biomass in the activated sludge system are discussed in Chapter 4.

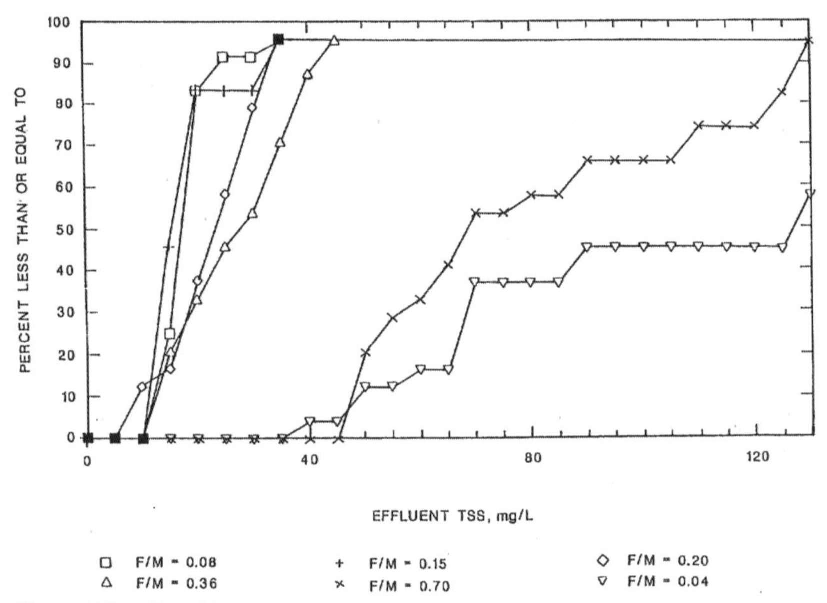

Figure 3.25. Effect of F/M on effluent suspended solids for an organic chemicals wastewater [4].

OXYGEN TRANSFER

Oxygen transfer principles and aeration are covered in detail in other references [1,2] and not presented herein. Aeration equipment technologies are also covered in these references and briefly discussed in Chapter 4.

REFERENCES

1. Eckenfelder, Jr., W. Wesley, Davis L. Ford and Andrew J. Englande, Jr. *Industrial Water Quality*. Fourth Edition. McGraw Hill (2008).
2. *Wastewater Engineering Treatment and Reuse*. International Edition. Metcalf and Eddy. Fourth Edition. 2004.
3. *Industrial Wastewater Management, Treatment and Disposal.* Third Edition. Water Environment Federation. McGraw Hill. 2005.
4. Eckenfelder, W.W. and Jack L. Musterman. *Activated Sludge Treatment of Industrial Wastewater*. Technomic. 1995.
5. Lester, J.C. 1987. *Heavy Metals in Wastewater and Sludge Processes*—Volume II, Treatment and Disposal. CRC Press.
6. Pitter, P. and J. Chudoba. 1990. *Biodegradability of Organic Substances in the Aquatic Environment.* CRC Press. Boca Raton.
7. Watkin, A. 1986. "Evaluation of Biological Rate Parameters and Inhibitory Effects in Activated Sludge." Ph.D. dissertation, Vanderbilt University.
8. Watkin, A. and W.W. Eckenfelder. 1988. "A Technique to Determine Un-steady State Inhibition in the Activated Sludge Process." *Water Science Technology, 21*:593–602.

9. Tabak, H.H., S.A. Quave, C.I. Mashni and E.F. Barth. 1981. "Biodegradability Studies with Organic Priority Pollutant Compounds." *J. Water Pollution Control Federation, 53*:1503.
10. Volskay, V.T. and P.L. Grady. 1988. "Toxicity of Selected RCRA Compounds to Activated Sludge Microorganisms." *J. Water Pollution Control Federation, 60*:10, 1850.
11. Philbrook, D.M. and C.P. Grady. 1985. "Evaluation of Biodegradation Kinetics for Priority Pollutants." Proc. 40th Industrial Waste Conference, Purdue University.
12. Larson, R.J. and S.L. Schaeffer. 1982. "A Rapid Method for Determining the Toxicity of Chemicals to Activated Sludge." *Water Research, 16*:675.
13. Daigger, G.T. and R.E. Roper. 1985. "The Relationship Between SVI and Activated Sludge Settling Characteristics." *J. Water Pollution Control Federation, 57*:859.

CHAPTER 4

Activated Sludge Process Technology Evolution and Trends

THE history and evolution of the use of activated sludge and the key process in its many configurations, modifications, technologies and advancements continue today. The advancements to the processes improve performance, reduce footprint, reduce energy costs and sludge production, as well as achieve a higher quality of water for more stringent permit limits or for recycle and reuse. This chapter presents a history of the evolution of activated sludge technologies and advancements and some of the key advantages and disadvantages of each of the technologies. The focus is on activated sludge suspended growth systems. Fixed film biological treatment systems such as trickling filters and rotating biological contractors (RBCs) are well covered in other references [1] and not discussed herein. Table 4.1 shows a summary of the history of technology advances described herein.

There are many technologies and technology providers in the business. The intent in this chapter is to show the more common ones that the authors have encountered and worked on in their careers. This chapter also includes a discussion of treating industrial wastewater in publicly owned treatment plants (POTWs) or municipal plants.

The evolution of the activated sludge technologies over the years has focused on the following improvements:

- Increasing biomass concentration and retaining biomass in the system
- Reducing footprint
- Increasing aeration and oxygen transfer
- Selecting and deselecting specific microorganisms to achieve better treatment efficiency and performance
- Protect more sensitive organisms such as nitrifiers from inhibitory constituents
- Improving treated effluent quality.

TABLE 4.1. History of Activated Sludge.

1913	Initial Research Work at Lawrence Experimental Station in Massachusetts
1914	Municipal wastewater research in England
1920s	First plant in United States in Massachusetts
1940s	Step feed plant in New York City
1945–1950	300 plants in United States
1950s	Aerated Lagoon
1960s	SBRs
1970s	PACT, Pure Oxygen, Deep Shaft
1980s	Thermophille Aerobic Treatment
1990s	MBRs, MBBRs, Selector
2000s	Annamox, Sharon and Nereda

PLUG-FLOW ACTIVATED SLUDGE

The plug-flow activated sludge process modification used in New York and other major cities and some industries uses long, narrow aeration basins to provide a mixing regime that approaches plug-flow conditions as shown in the picture in Figure 4.1 and diagram in Figure 4.2. A plug-flow regime promotes the growth of a flocculent and well-settling sludge by introducing influent wastewater and return sludge at the head-end of the basin. This provides a high substrate gradient that promotes growth of floc forming, rather than

Figure 4.1. Plug flow activated sludge [2].

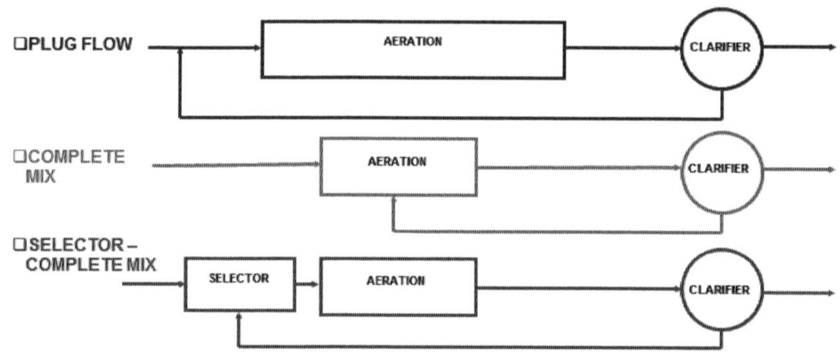

Figure 4.2. Types of activated sludge processes.

filamentous biomass. If the wastewater contains toxic or bioinhibitory organics, however, they must be removed or equalized prior to entering the head-end of the aeration basin since the flow pattern provides negligible dilution in the aeration basin itself. The oxygen utilization rate is high at the beginning of the basin and decreases with aeration time. Under normal operating conditions, the mixed liquor oxygen utilization rate approaches the endogenous level toward the end of the aeration basin.

Modification of the way in which wastewater and return sludge are brought into contact in a plug-flow system can have a number of benefits. Provision of a separate zone at the inlet, with a volume of about 15 percent of the total aeration volume and a low-energy subsurface mechanical mixer, can achieve controlled anoxic conditions. These conditions generally promote good floc formation and control of filamentous growth. Step feed is also used to introduce the wastewater at multiple points. In cases where nitrification occurs, recycle of nitrified mixed liquor (and NO_3) from the end of the aeration basin to the anoxic zone at the head-end can achieve significant denitrification and remove some of the influent BOD without oxygen supply (see MLE Process).

COMPLETE MIX-ACTIVATED SLUDGE

In a complete mix-activated sludge process, as shown in Figure 4.2, the wastewater and the return sludge are introduced into the aeration basin at a few points to facilitate their rapid blending with the basin contents. The objective is to maximize equalization of the influent load within the aeration basin. This process is particularly applicable to industrial wastewaters that contain toxic/bioinhibitory substances or that have highly variable loading patterns. Furthermore, wastewaters with variable pH are neutralized by the basin contents and biological activity. Another advantage of a complete mix process is that the oxygen uptake rate is equalized throughout the basin, thus permitting uniform spacing of the aeration equipment.

Complete mix-activated sludge should not be employed when treating readily degradable wastewaters unless a biological selector is installed in front of the aeration basin in order to maintain good sludge settling characteristics. Performance data for various industrial wastewaters were presented in Table 4.2. A biological selector upfront of the aeration basin encourages the growth of floc-forming organisms and maintains sludge quality control. This is particularly important when treating readily degradable wastewaters.

EXTENDED AERATION

The extended aeration process utilizes a longer hydraulic retention time (18 to 24 hr) and a low *F/M* (high SRT). This results in minimal sludge production but high oxygen requirements per pound of BOD removed. The *F/M* typically will vary from 0.05 to 0.15/day (SRT of 20 to 40 days), and MLSS concentrations will range from 3,000 to 5,000 mg/L. The process may be operated in either a complete mix or plug-flow mixing regime. It is primarily used in smaller industries where simplicity of operation and low sludge production are important. It is also applicable for treatment of poorly degradable organics that require high SRT to satisfy discharge limits.

AERATED LAGOON

Aerated lagoons have been used for years to treat industrial wastewater including pulp and paper and textiles. They are still in use today and are typically a completely mixed lagoon with surface or diffused aerators and no final clarifier with recycle of sludge back to the aeration basin. Figure 4.3 shows a simplified schematic showing the aerated lagoon followed by a facultative lagoon and a settling lagoon. Aerated-stabilization basins (ASBs) which are commonly used in the pulp and paper industry are aerated lagoons upfront with settling basins at one end of the lagoon or in a separate lagoon as in Figure 4.3.

OXIDATION DITCH

There are a number of oxidation ditch systems available. In these systems, it is necessary to match basin geometry and aerator performance in order to yield an adequate midchannel velocity (>1 fps) for mixed liquor solids transport. The key design factor in these systems relate to the type of aeration that is provided. This technology is particularly applicable to those cases where both BOD and nitrogen removal are desired since both reactions can be achieved in the same basin by alternating aerobic and anoxic zones. Figure 4.4 shows a picture and diagram of the oxidation ditch.

TABLE 4.2. CMAS Treatment Performance for Selected Industrial Wastewaters [3].

Wastewater	Influent		Effluent			F/M						
	BOD, mg/L	COD, mg/L	BOD, mg/L	COD, mg/L	T, °C	BOD, d^{-1}	COD, d^{-1}	SRT, *d*	MVLSS, mg/L	HRT, *d*	SVI, ml/g	ZSV, ft/h
Pharmaceutical	2950	5840	65	712	10.4	0.11	0.19		4970	5.4		
	3290	5780	23	561	20.8	0.11	0.18		5540	5.4		
Coke and by-products chemical plant	1880	1950	65	263		0.18	0.21		2430	4.1	42.4	26
Diversified chemical industry	725	1487	6	257	21	0.41	0.71		2874	0.61	119	4.45
Tannery	1020	2720	31	213	21	0.18	0.45	16	1900	3		
	1160	4360	54	561	21	0.15	0.49	20	2650	3		
Alkylamine manufacturing	893	1289	12	47		0.146	0.21		1977	3.1	133	4.2
ABS	1070	4560	68	510	335	0.24	0.94	6	2930	1.5	23	28.7
Viscose rayon	478	904	36	215		0.30	0.47		2759	0.57	117	4.7
Polyester and nylon fibers	207	543	10	107	131	0.18	0.40		1689	0.664	116	7.9
	208	559	4	71	22.4	0.20	0.48		1433	0.712	144	8.6
Protein processing	3178	5355	10	362	10	0.054	0.08		2818	21	180	2.9
	3178	5355	5.3	245	26.2	0.100	0.16		2451	12.7	215	2.7
Propylene oxide	532	1124	49	289	20	0.20	0.31		2969	1	51	12.5
	645	1085	99	346	37	0.19	0.25		2491	1.4	32	3.7
Paper mill	375	692	8	79	9.2	0.111	0.19	18.9	1414	2.38	63	22
	380	686	7	75	23.3	0.277	0.45	5.2	748	1.83	504	10
Vegetable oil	3474	6302	76	332		0.57	1.00		1740	3.5	49.2	30
Organic chemicals	453	1097	3	178	20.3	0.10	0.21		2160	2.02	111	6.9

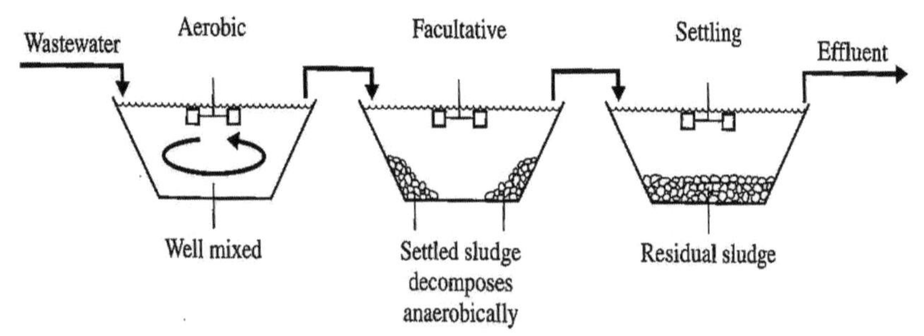

Figure 4.3. Aerated lagoon.

SEQUENCING BATCH REACTOR

In the Sequencing Batch Reactor (SBR) process, a single vessel provides both the biological oxidation and settling processes that are normally associated with multi-tank conventional activated sludge treatment without a separate clarifier. The SBR process can provide complete nitrification and substantial denitrification with few facility modifications by changing the timing and duration of the aeration and fill-draw cycles.

The anoxic or selector part of the overall SBR cycle of 6 to 12 hours, provides for exposure of biomass to elevated substrate concentrations that favor the growth of floc-forming microorganisms. The oxidation reduction potential

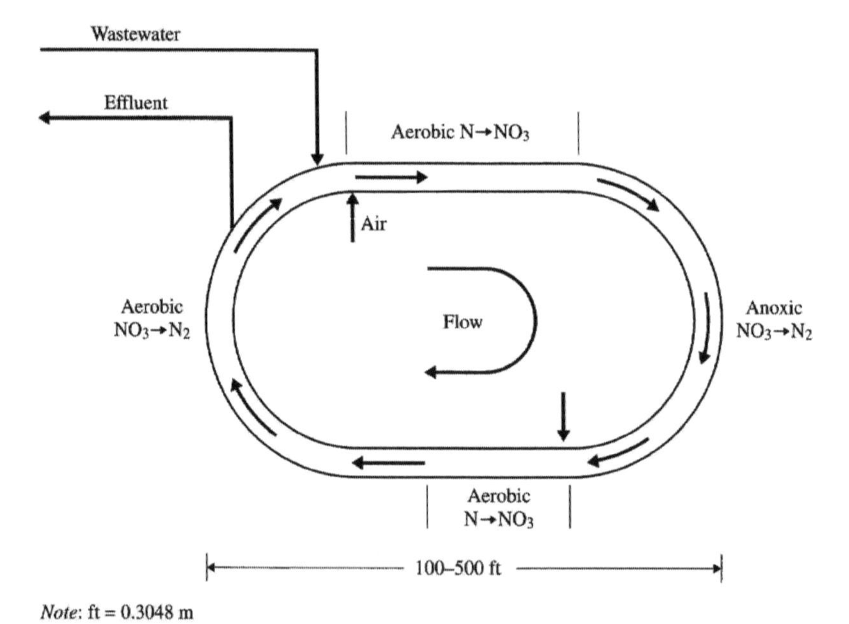

Figure 4.4. Oxidation ditch with nitrification and denitrification.

(ORP) in this zone can be controlled by the aeration intensity, cycle time, and organic loading rate as needed to favor bioselective mechanisms.

Figure 4.5 shows the picture of an SBR with two compartments: one is in aeration and one is settling and decant mode with the aeration turned off. Most facilities incorporate multiple basins operated in parallel. The basic sequences in a cycle of operation are: aerated fill, anoxic, aeration, settle, decant and idle. A moving weir is used to withdraw an upper volume of supernatant (treated effluent) from the reactor and has adjustments to permit changes to the rate and depth of supernatant removal. Operating cycles for industrial wastewaters vary from 6 to 96 hr, depending on the hydraulic-organic loading relationship, with 2 hr typically used for the settle-surface skim sequence. Plants can be designed on an average *F/M* ratio of 0.05 to 0.20 lb BOD/lb MLVSS-day, depending on the quality of effluent required. A design example for an SBR treatment of pharmaceutical wastewater for a 6-hour cycle time is presented in Chapter 7.

A design example for a SBR process for removal of IPA and acetone from a pharmaceutical wastewater is presented in Chapter 7.

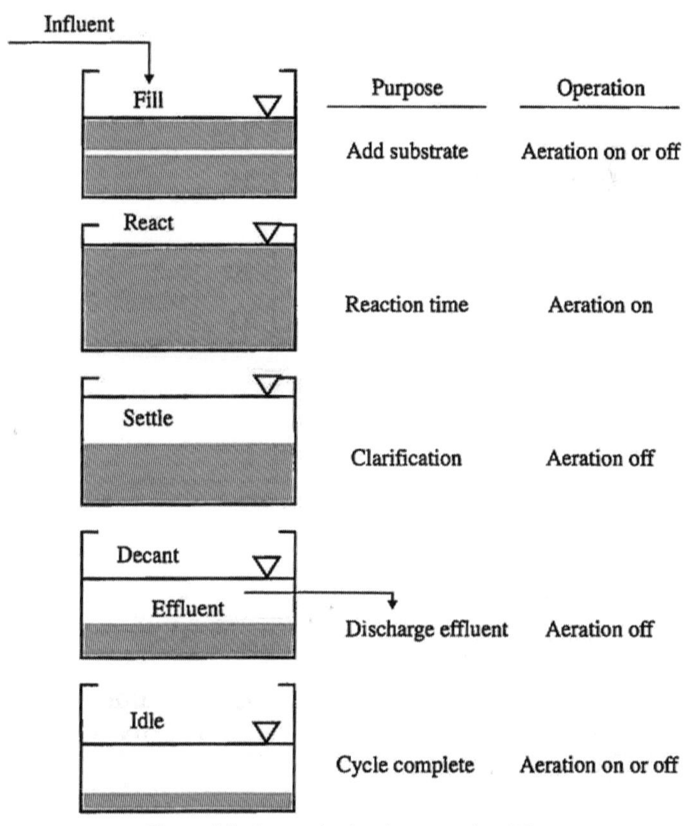

Figure 4.5. Sequencing batch reactor plant [3].

HIGH-PURITY OXYGEN SYSTEM

The high-purity oxygen system shown in Figure 4.6 is a series of well-mixed reactors employing concurrent gas-liquid contact in a covered aeration tank. Feed wastewater, recycle sludge, and oxygen gas are introduced into the first stage. The oxygen gas is fed at low pressure (approximately 1.5 in H_2O). Two gas-liquid contact systems can be employed: submerged turbine aeration, and surface aeration. With turbine aeration, recirculating gas blowers pump the gas through a shaft to a rotating sparger. The pumping action of the impeller on the same shaft as the sparger promotes adequate liquid mixing and yields relatively long residence times for the dispersed oxygen bubbles. Gas is recirculated within a stage at a rate that is usually higher than the rate of gas flow between stages. A slight pressure drop occurs from stage to stage to prevent gas back-mixing. Since the relative liquid mixing and oxygen transfer requirements vary from stage to stage, each stage is equipped with an independent mixer-compressor combination designed to provide the required level of mixing and oxygenation.

The gas-liquid contact mechanism provided by surface aerators eliminates the need for gas recirculating compressors and associated piping. The mixing intensity required to maintain the sludge in suspension is provided by a low-speed, low-shear impeller. Oxygen gas is automatically fed to either system on a demand basis (based on headspace gas pressure) with the entire unit operating, in effect, as a respirometer. As the organic loading rate increases, the oxygen gas pressure decreases, resulting in an automatic increase in feed-oxygen flow.

Due to the high mixed liquor solids maintained in the oxygen system, the major portion of soluble BOD removal and, thus, the highest oxygen demand occur in the first stage, which requires the highest mixer and compressor horsepower. The subsequent stages are then utilized to stabilize a sludge that has a progressively decreasing oxygen demand. Effluent mixed liquor from the system is settled, and the clarifier underflow is returned to the first stage for blending with the feed. The exhaust gas from the final stage is vented to the atmosphere. The system normally operates with a vent-gas composition of 30 to 50 percent oxygen. Due to the net transfer of gas to the liquid, the vent-gas flow rate will be only 10 to 20 percent of the gas feed rate. Based upon economic considerations, about 90 percent oxygen utilization is desired.

Two basic oxygen generation processes are employed: a traditional cryogenic air separation process for large installations (greater than 20 tpd) and a pressure swing adsorption (PSA) system for smaller installations. With larger installations, deep tank construction with submerged turbine aeration is preferable, while a surface aerator-PSA combination is the most cost-effective for smaller plants. The power requirements for the surface and turbine aeration equipment vary from 0.08 to 0.14 horsepower per thousand gallons, depend-

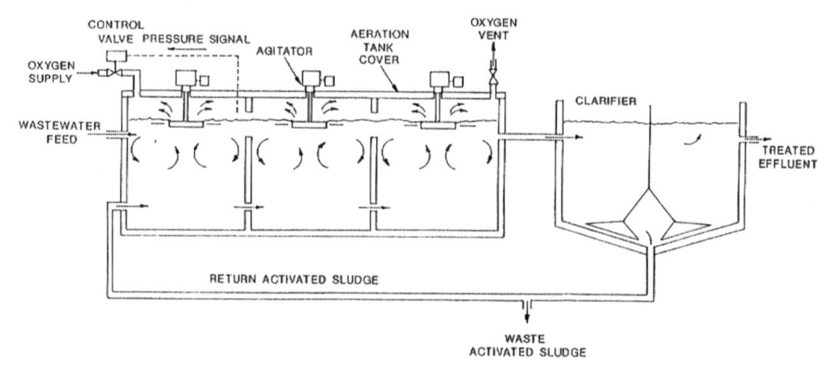

Figure 4.6. Flowsheet for high-purity oxygen process.

ing on the waste strength, mixing requirement, feed oxygen purity, and the capacity of the aeration equipment. The oxygenation systems are typically designed to maintain 6 mg/L dissolved oxygen in the mixed liquor at peak load conditions. Liquid oxygen storage is provided for backup purposes with the same supply capacity as the installed plant. It is therefore possible to double the feed-oxygen flow to the aeration tank if needed to satisfy unusual organic loading events. This results in an increased gas phase oxygen partial pressure and increased oxygen transfer, but reduced oxygen utilization. This is not an economical mode of operation over extended time periods.

The maintenance of a fully aerobic floc will maximize the endogenous rate coefficient (b) and thereby minimize the sludge production under moderate to high F/M loading conditions. The sludge settling rate will also be at a maximum under fully aerobic conditions. These conditions are promoted under the high mixed liquor dissolved oxygen operating levels that are available with the pure oxygen-activated sludge process, especially for high-strength wastewaters. The Hopewell, VA POTW, which treats a high industrial wastewater contribution, uses this process.

DEEP SHAFT-ACTIVATED SLUDGE

The deep shaft-activated sludge process shown in Figure 4.7 operates at a F/M of l/day to 2/day (BOD basis) using a mixing energy level of 800 to 1,500 hp/MG of aeration basin volume. The shaft depth varies from 150 to 400 ft. The operating mixed liquor dissolved oxygen concentration varies from 10 mg/L to 20 mg/L since the increasing shaft depth increases the saturation concentration. The MLSS levels vary from 8,000 mg/L to > 12,000 mg/L. Solids-liquid separation is provided by dissolved air flotation at high MLSS concentrations (> 10,000 mg/L) and by vacuum degasification and conventional gravity clarification at lower MLSS levels.

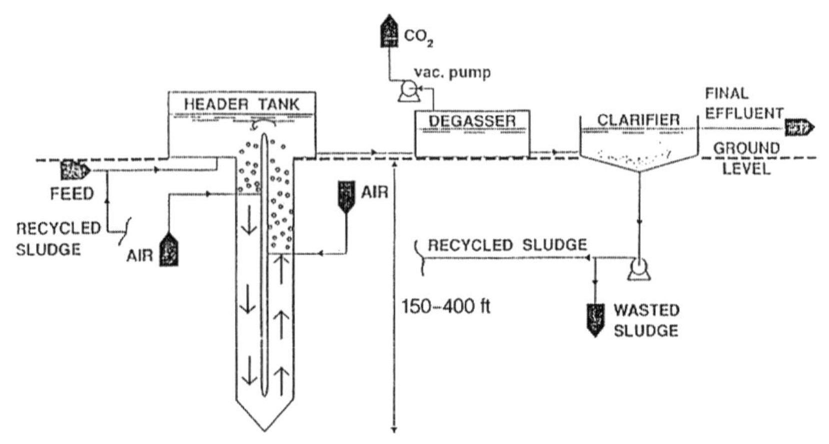

Figure 4.7. Flowsheet for deep-activated sludge process [2].

BIOHOCH® PROCESS

The BIOHOCH® reactor shown in Figure 4.8 consists of an aeration section divided by a perforated plate into a lower and an upper zone, and a cone-shaped final clarifier surrounding the aeration section. The air is supplied to the reactor by means of radial flow jets installed at the reactor bottom.

The untreated wastewater is pumped into the reactor through the radial flow jets or through separate pipes. Turbulence in the lower zone is sufficient to provide completely mixed conditions. In the upper chamber, the stabilizing and degassing zone, bubbles adhering to tile activated sludge are removed, since they impede sedimentation in the final clarifier. The depth of the aeration zone

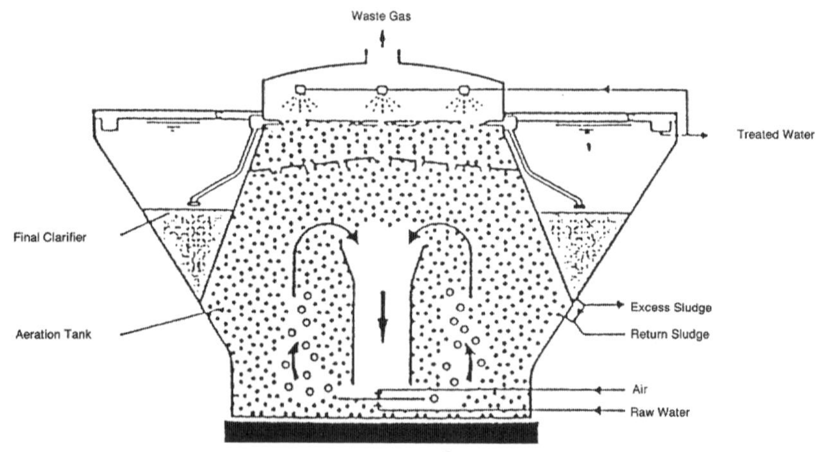

Figure 4.8. Flow scheme of a BIOHOCH® reactor (courtesy of Hoechst) [2].

TABLE 4.3. Treatment of an Organic Chemicals Wastewater in the BIOHOCH® Reactor [2].

Parameter	Value
Flow rate	0.63 MGD
Influent BOD	5,000 mg/L
Effluent BOD	40 mg/L
Influent COD	6,000 mg/L
Effluent COD	750 mg/L
F/M (BOD basis)	0.43 lb/lb MLSS-day
MLSS	3,500 mg/L
SRT	5.4 days
HRT	80 hr
Temperature	95°F

is approximately 65 ft. Performance data for treatment of an organic chemicals wastewater are shown in Table 4.3.

The BIOHOCH® reactor can be combined with a modified dissolved air flotation unit in place of a conventional secondary clarifier. This permits a recycle solids of 40 g/L and a resulting MLSS of 10 g/L. Reported performance data indicated an effluent TSS of less than 20 mg/L.

POWDERED-ACTIVATED CARBON (PACT)

The PACT® process is the addition of powdered-activated carbon to the activated sludge process. Powdered-activated carbon is applied to the activated sludge process when conventional biological oxidation cannot meet effluent requirements. This usually involves the removal of color, nondegradable COD or roc, effluent aquatic toxicity, or bioinhibition of heterotrophic or nitrifying organisms. A flow schematic of a PACT® system is shown in Figure 4.9. Powdered-activated carbon is mixed with the influent wastewater or fed directly to the aeration basin. The carbon-biosludge mixture is settled and the sludge recycled in the same manner as the conventional activated sludge process. The waste-activated sludge contains carbon, biological sludge, and residual influent solids.

The PACT process offers an advantage for upgrading the performance of existing activated sludge treatment facilities since it can usually be integrated into the facility at lower capital cost than a fixed-bed granular carbon adsorption system. Since the addition of PAC enhances sludge settleability, secondary clarifier area requirements based on conventional hydraulic and solids flux loading rates will usually be adequate, even at high carbon dosages. One problem associated with the application of PAC is that the resulting carbon-biomass sludge is abrasive. Appropriate materials of construction must, therefore, be provided for pumps, tankage, clarifier rake mechanisms, and sludge handling equipment. The high density of the resulting sludge will also require increased torque limits on the final clarifier rake mechanism and subsequent sludge processing equipment.

The PACT mixed liquor will be a mixture of biological floc, carbon particles, and biomass growing on the carbon. Both degradable and nondegradable organics are adsorbed on the carbon, but it is assumed that, at high SRT, most of the degradable organics will be biologically oxidized. In general, the adsorption capacity of the carbon in the PAC sludge will be higher than that predicted by a batch adsorption isotherm. This is illustrated in Figure 4.10. Typical process design criteria are SRT = 5 to 20 days, MLSS = 15 to 20 g/L, and PAC feed rates of 50 to 3,000 mg/L. Actual operating conditions, however, depend on effluent requirements and the purpose of PAC addition. For this reason, continuous flow treatment studies should be conducted to develop process design and operating criteria for PACT applications. The concentration of PAC in the mixed liquor (PAC_{ML}) varies with PAC addition rate (Dose) and can be determined by Equation (4.1).

$$PAC_{ML} = \frac{\text{Dose } \theta_c}{t} \tag{4.1}$$

The results of PACT treatment of an organic chemicals wastewater are summarized in Table 4.4. They demonstrate that varying PAC dosages were required to achieve effluent requirements for STOC and color. The PAC dosage, however, had no effect on acute toxicity or SBOD of the effluent. Furthermore,

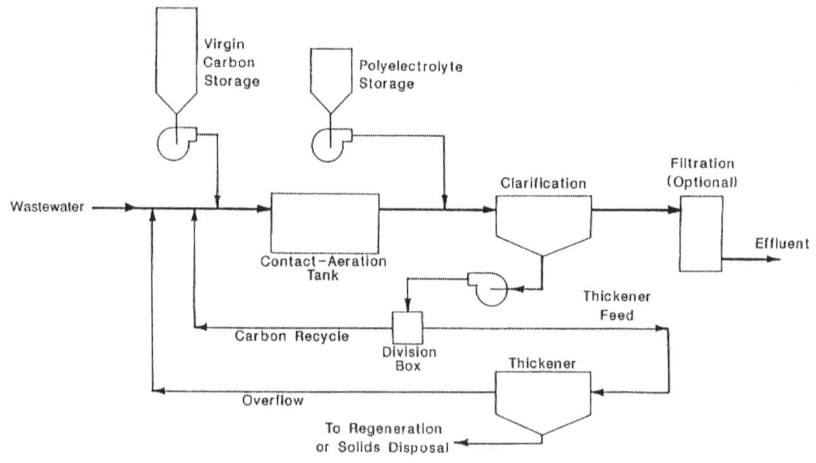

Figure 4.9. Flow diagram for PACT® wastewater system.

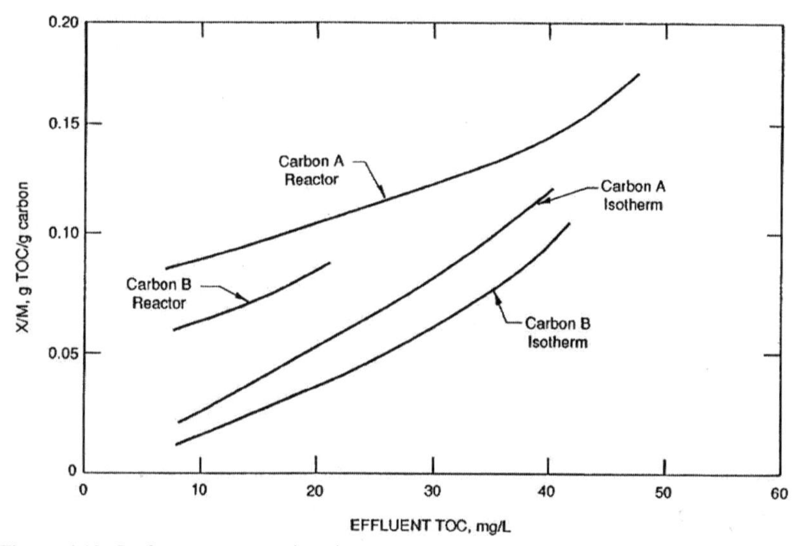

Figure 4.10. Performance comparison between continuous PAC reactors and their adsorption isotherms.

tertiary treatment with carbon (following conventional activated sludge) provided slightly better effluent STOC and color quality than PACT treatment at the same carbon utilization rate. TOC removal with PACT for several case studies is shown in Table 4.5.

Nondegradable TOC or COD is removed by PAC, but the removal efficiency depends on the chemical characteristics of the TOC and COD material. Much of the color present in textile, pulp and paper, and dyestuff wastewaters is nonbiodegradable but absorbable on activated carbon.

TABLE 4.4. Effects of Varying PAC Dosages on Activated Sludge Treatment on an Organic Chemicals Wastewater [2].

		Effluent Quality at Indicated PAC Dosage (mg/L)				
Parameter	**Influent**	**0**	**250**	**500**	**1,000**	**Tertiary PAC Contractor at 1,000 mg/L PAC[(a)]**
SBOD, mg/L	460	21	26	23	20	NA[(b)]
STOC, mg/L	380	146	128	121	91	70
Color, APHA	1,130	1,140	750	540	300	240
TSS, mg/L	NA	20	54	27	25	NA
SOUR, mg/g-hr	NA	4.9	6.3	5.4	4.2	NA
48 hr LC_{50}[(c)], %	NA	19	32	32	33	32

[(a)]Treating effluent from activated sludge reactor without PAC addition.
[(b)]Not analyzed.
[(c)]With *Daphnia pulex*.

TABLE 4.5. Operating Characteristics for PACT Process Case Studies [2].

	Case Study No.				
Operating Parameter	**1[(a)]**	**2[(b)]**	**3[(b)]**	**4[(b)]**	**5[(b)]**
Aeration system					
HRT, days	0.32	0.75	2.3	3.8	4.2
SRT, days	54	40	5.8	20	19.3
MLSS, g/L	34.8	10–12	–	–	–
PAC Dosage, mg/L	114	170	2,270	850	1,140
TOC, Influent, mg/L	174	–	2,470	2,330	2,490
TOC, Removal, %	81	–	85.3	98.9	98.4

[(a)]Zimpro, 1984.
[(b)]Lankford, P. and W. Eckenfelder. 1990. *Toxicity Reduction in Industrial Effluents.* Van Nostrand Reinhold, New York

The concentrations of certain heavy metals are effectively reduced in the PACT process. Metals removal may occur by adsorption of an organic that has complexed the metal, by surface precipitation with sulfide occurring on the carbon due to its high sulfur content, or by co-precipitation with the biological floc. Results of metals removal by PACT from an organic chemicals wastewater are shown in Figure 4.11.

Another potential application of the PACT technology is for reduction of the aquatic toxicity of a biologically treated effluent. No consistent relationship, however, has been established between reduction of aquatic toxicity and reduction of other parameters, such as TOC, NH_3-N, or metals. Toxicity reduc-

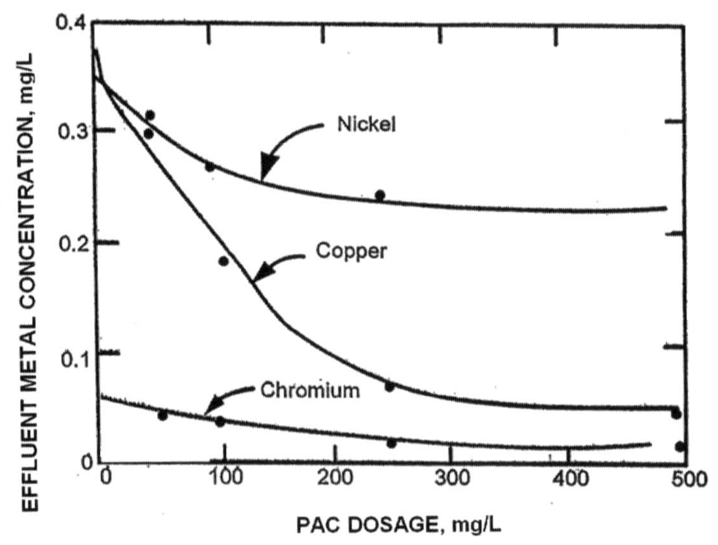

Figure 4.11. Metals removal from organic chemicals wastewater using PAC [2].

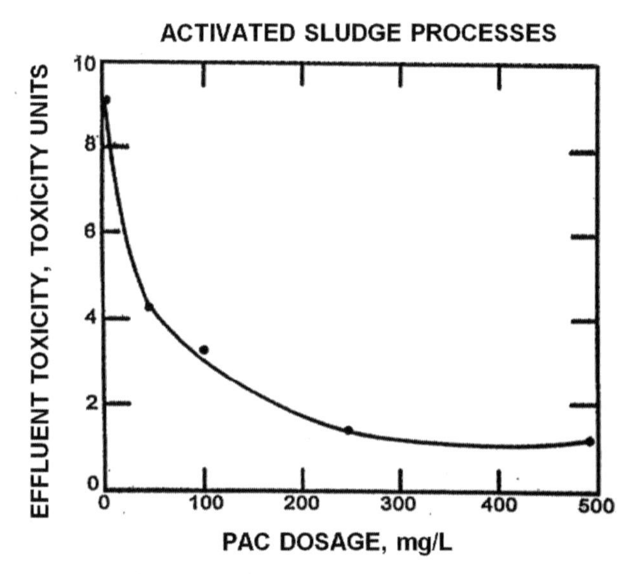

Figure 4.12. Toxicity reduction in an organic chemicals wastewater using PAC [2].

tion results from PACT treatment of an organic chemicals effluent are shown in Figure 4.12.

As previously discussed, biological nitrification is sensitive to the presence of heavy metals and certain organic compounds. The application of PAC can reduce or eliminate this inhibition by reducing the "free" concentration of the toxic agent through adsorption and subsequent biodegradation. Results reported by Briddle *et al.* [4] on a coke plant effluent are summarized in Table 4.6. The application of PAC to a chemical plant wastewater to enhance nitrification is shown in Figure 4.13.

The addition of PAC to the activated sludge process also reduces the loss of VOCs to the gas phase. The mechanism for VOC removal is adsorption and enhanced biodegradation (on the carbon). This removal strategy can be significant when considering control of VOC from the activated sludge process.

TABLE 4.6. Effect of PAC on Nitrification of Coke Plant Wastewaters [4].

		Effluent Characteristics				
PAC Dosage (mg/L)	**SRT (days)**	**TOC (mg/L)**	**TKN (mg/L)**	**NH_3-N (mg/L)**	**NO_2-N (mg/L)**	**NO_3-N (mg/L)**
0	40	31	72.0	68	4.0	0
33	40	20	6.3	1	1.0	9.0
50	40	26	6.4	1	1.0	13.0

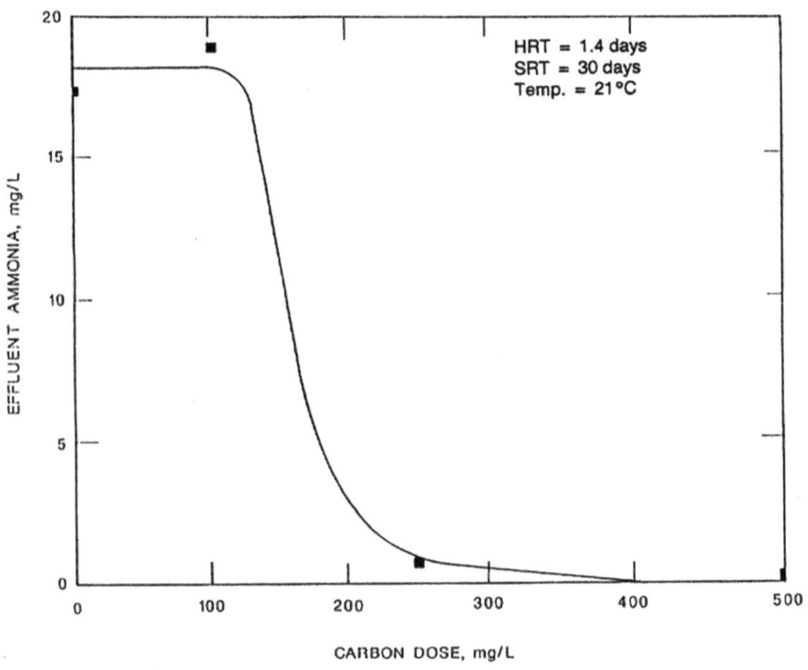

Figure 4.13. Effect of carbon dose on effluent ammonia concentration [2].

MODIFIED LUDZACK ETTINGER (MLE) PROCESS

Figure 4.14 shows the MLE activated sludge treatment configuration developed in 1973 by Barnard as an improvement on the original process in 1962 [1] with upfront anoxic selector. This technology is being used more now versus the more traditional activated sludge and sequencing batch reactor (SBR) for nitrogen removal with recycle of the MLSS back to the anoxic zone for nitrate removal. There is some synergy in goals as nutrient removal plants with increased sludge age should achieve increased removal of the microconstituents that are biodegradable as discussed in Chapter 6. The MLE process uses the nitrate from recycle of MLSS instead of oxygen in the anoxic selector to remove the COD and BOD thereby reducing the oxygen requirements and energy costs in the aeration basin. The denitrification process in the anoxic selector also reduces the alkalinity required for nitrification.

MEMBRANE BIOREACTOR (MBRs)

The membrane biological reactor (MBRs), which is shown in Figure 4.15, is one of the key advances over the years for the activated sludge process.

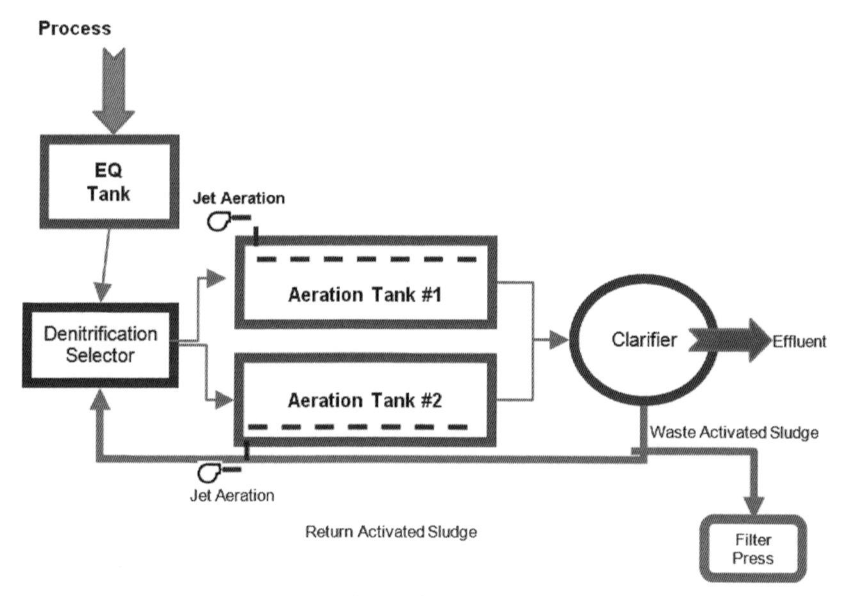

Figure 4.14. Activated sludge with anoxic selector.

The MBR process is a suspended growth aerobic biological treatment system featuring an aeration basin followed by membrane filtration. The membranes can be internal in the aeration basin or in a separate tank or external as in Figure 4.15. The MBR is essentially an activated sludge process in which the secondary clarifier is replaced by membranes. The membrane filtration following the bioreactor provides a physical barrier to retain a very high level of MLSS allowing the MBR process to operate in smaller footprints than conventional systems. The types of membranes that could be used include microfiltration, ultrafiltration and nanofiltration. The specific membrane application would depend on the level of removals required for suspended solids, turbidity or particulates. Operating at a low food to micro organism

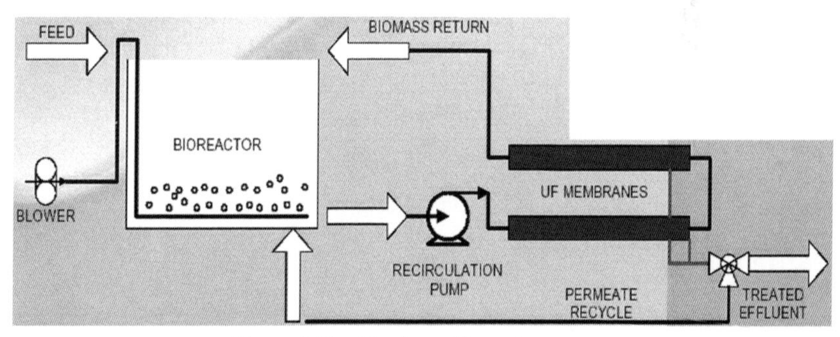

Figure 4.15. Membrane bioreactor process.

ratios (loadings) allows the MBR to accept high volumetric and organic loadings while operating at low hydraulic retention times. The higher SRTs in the MBR would result in lower biological sludge production. The high level of treatment in the bioreactor along with suspended solids retention by the membranes produces a high quality effluent with very low levels of BOD and suspended solids. Biological treatment using MBRs has a smaller footprint compared to conventional activated sludge. The MBR technology is typically used for either plant upgrades or new plants in which water recycled and reused is being considered. MBR followed by RO is a good solution on water reuse projects.

There are over ten MBR technology providers in the business and a number of companies that provide the membranes. There are three types of membrane configurations. There are membrane cassettes immersed right in the aeration basin mixed liquor. There are external membranes outside the aeration basin in either a horizontal position or in a vertical position. The vertical position like the immersed uses air scouring to clean the membranes and minimize build-up and fouling. The external horizontal membranes do not use air scour to rely on the cross flow velocity for membrane cleaning. There are over 120 spiral membranes and over 120 flat plate MBRs in operation.

The two traditional configurations of membrane bioreactor systems. The immersed membrane configuration has been more prevalent throughout North America because of the lower cost of operation when compared to traditional external cross flow systems. The lower operating cost capability of the immersed systems stems from the ability to hydraulically clean the membrane with a two phase flow combination of air and mixed-liquor. The two phase flow provides a shearing effect on the membrane surface controlling the amount of sludge cake deposit on the membrane surface, helping to maintain membrane permeability. It has also been demonstrated by Kang [5] that an MBR system performance improves 20–60% when a two-phase flow is utilized rather than liquid only because of the additional turbulence created. The flow path configuration of the submerged membranes is from the outside of the membrane surface to the inside.

Traditionally external MBR systems have arranged the tubular membrane modules in a horizontal fashion and rely on hydraulically pumped mixed-liquor to create a cross-flow shearing force on the membrane surface layer. The flow path of the water flows from the inside of the membrane tube to the outside. Since the membranes are aligned horizontally the use of air to ensure membrane performance is not an option, requiring a significant amount of horsepower (hp). The advantage the external system has is additional tank(s) to house the membrane are not required and operator exposure to hazardous mixed-liquor is limited.

Even though the liquid/solid separation configuration of the external and immersed system is different, each has positive design aspects. Combining

TABLE 4.7. Annual Electrical Consumption of Airlift MBR—Venezuela Brewery [6].

Equipment	Operating HP	Operational Hours per Day	kW-hr/year
Membrane Feed Pump	6.1	24	79,226
Blower—Airlift	17.1	24	111,779
Membrane Permeate Pump	1.1	23.2	13,552
BW Pump	17.6	0.8	3,837
		Total kW-hr/year	**208,365**
		Total kW-hr/1,000 gal treated	**0.95**

the benefits of each can be found in the more recently developed hybrid Airlift™ membrane bioreactor system (Sparks, Turner) [6]. The hybrid Airlift system is an external tubular membrane that arranges the membrane modules vertically. It then combines activated sludge with compressed air utilizing a coarse bubble diffuser below the membrane module tubes. The air makes the mixed-liquor more buoyant, forcing it up through the membrane tubes. The volume of fluid is pumped at a controlled rate ensuring that enough scouring velocity is allotted to keep the membrane at a high level of performance. The use of the airlift principle allows for an effective scour of the membrane surface along with reducing the energy draw required to operate the system.

The most significant advantage of the hybrid Airlift MBR system is associated with the power consumption associated with the operation of membrane system. As shown in Table 4.7, the Airlift system consumes 0.95 kW-hr for every 1,000 gallons treated at the design flow capacity of the brewery wastewater treatment system. This equates to a reduction in electrical consumption of 27%, 46% and 93% when compared to immersed-hollow fiber, immersed-

TABLE 4.8. Membrane Operating System Power Consumption—HF, FS and Cross Flow [6].

Immersed-Hollow Fiber			
Equipment	**Operating HP**	**Operational Hours per Day**	**kW-hr/year**
RAS Pumps	2.1	24	27,651
Blower—Airlift	37.1	24	242,319
Membrane Permeate Pump	1.1	24	13,989
		Total kW-hr/year	**283,959**
		Total kW-hr/1,000 gal treated	**1.29**

flat sheet and cross flow-tubular, respectfully. The electrical consumption of the various MBR systems is presented in Table 4.8.

The disadvantage of MBRs principally revolves around the operating cost associated with maintaining the performance of the membranes. There are four different types of MBR configurations currently offered in the market. Each of the configurations is capable of achieving similar effluent, provided the activated sludge system is designed properly. The variance in MBRs is associated with the membrane configuration and method to sustain performance of the membranes. In the case of the 602,000 gallon per day Venezuelan brewery MBR it has been demonstrated that Airlift MBR system utilizes 27% less energy for the membrane operating system when compared to immersed hollow fiber system, 46% less energy than immersed flat sheet, and 93% less than cross flow tubular MBRs if operated under the same conditions.

Some of the key questions and issues to be evaluated in selecting the type of membrane system (i.e., flat plate versus spiral wound) are as follows:

- Upfront screening facility needs
- System complexity
- Total anoxic and aeration basin volume
- Overall system foot print
- Maximum operating MLSS
- Cleaning frequency
- Cleaning facilities and chemicals needed
- CAPEX costs
- OPEX costs
- Membrane life
- Membrane replacement costs
- Potential for fouling
- Oxygen requirements
- Operating horsepower

MOVING BED BIOREACTOR (MBBR)

This technology shown in Figure 4.16 is essentially the activated sludge process with the addition of plastic media in the aeration basin. The plastic media provides significant surface areas for biomass attachment to create a combined suspended and attached growth system. Some of MBBRs systems, such as a pharmaceutical company in Puerto Rico, do not need a clarifier to meet TSS permit limits for discharge to the POTW. The Integrated Fixed Film Activated Sludge (IFAS) process has a clarifier or DAF to reduce the effluent TSS concentrations and recycle return sludge to the aeration basin.

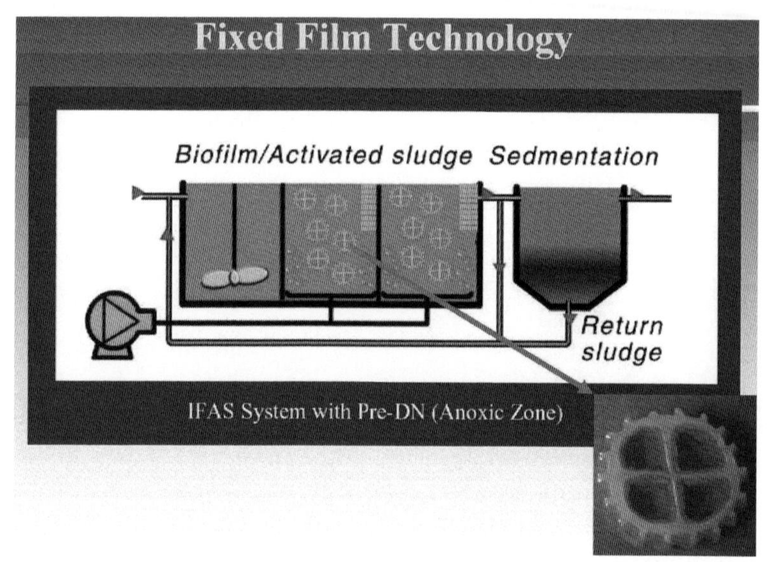

Figure 4.16. Fixed film technology.

BioMag AND CoMag

BioMag and CoMag Systems from Siemens are next generation wastewater and water solutions that increase the capacity and contaminant removal capabilities of primary, secondary, and tertiary treatment systems. Siemens purchased the technologies through the acquisition of Cambridge Water Technologies in early 2012. The technologies make the fundamental process of clarification far faster, far more reliable, and far more compact than alternate technologies. They give the operator far more control over effluent values, and a far more robust plant. This ground-breaking technology infuses magnetite as a weighting agent into chemical or biological floc, resulting in rapid and enhanced settling and clarification. The technology can be cost effectively applied to both new or existing treatment plants, and greatly reduces footprint and tankage. The process creates much more concentrated suspended solids, and a more dense and stable sludge blanket that makes plants less prone to load and flow variations, and delivers superior nutrient and solids removal. Magnetite is efficiently recovered and recycled.

The BioMag Ballasted Biological Treatment System enables an existing plant to double or triple its biological treatment capacity and achieve Enhanced Nutrient Removal limits all within existing tanks. BioMag routinely delivers:

- TN < 3.0 mg/L
- TP < 0.2 mg/L
- Turbidity < 1 NTU

The CoMag Particulate Removal System can produce effluent nearly equivalent to ultra-filtration at capital and operating costs competitive with conventional alternatives. It is a perfect technology for tertiary treatment and CSO and storm water management. CoMag routinely delivers:

- TP < 0.05 mg/L
- TSS < 2.0 mg/L
- Turbidity < 1 NTU

BioMag and CoMag Systems are gaining rapid acceptance by municipalities and engineers.

- BioMag Systems: 3 operating plants; 12 more in design
- CoMag Systems: 8 operating plants; 6 more in design

NEREDA® PROCESS

Nereda® is an innovative, advanced biological wastewater treatment technology that purifies water using the unique features of aerobic granular biomass. Unlike conventional processes, the purifying bacteria concentrate in compact granules, with excellent settling properties. As a result of the large variety of biological processes that simultaneously takes place in the granular biomass, Nereda® is capable of meeting stringent effluent quality requirements. Extensive biological phosphorus and nitrogen reduction is an intrinsic attribute of this technology, resulting in chemical free operation.

These unique process features translate into compact, energy saving and easy to operate Nereda® installations for both municipal and industrial wastewater treatment. Nereda® presents attractive new solutions for greenfield and brownfield sites and retrofitting or extending conventional activated sludge plants. The technology is also highly recommended for performance and capacity upgrades of existing SBR facilities.

Nereda® enables extensive biological treatment in a compact and simple design. Reduced plant costs are achieved through:

- High energy efficiency of the process
- Less mechanical equipment than conventional processes
- Concentrated biomass, reducing tank volume
- Reduced direct plant costs
- Reduced O&M costs

THERMOPHILIC (AFC℠) PROCESS

The AFC℠ technology treats waste streams containing high strength organic compounds and organic sludges [4]. AFC^{SM}'s main attribute is its ability to convert high strength and potentially toxic organics, both dissolved and suspended particulate, to carbon dioxide and water. AFC^{SM} eliminates the potential long-term liability risks and minimizes the costs associated with off-site disposal because it is a complete oxidation option. The AFC^{SM} process is a combination of single-stage thermophilic aerobic biological treatment, side stream chemical treatment, and efficient solids separation. Waste materials are fed to a self-heating thermophilic bioreactor for treatment. Effluent from this reactor is conveyed to a solids separator. Ultrafiltration is the preferred solids separation process for full-scale AFC^{SM} systems. A portion of the concentrated solids is returned directly to the thermophilic reactor, while the remaining portion goes to chemical treatment. The chemical treatment step partially solubilizes the excess biosolids before they are returned to the thermophilic reactor for further digestion and destruction.

INTEGRAL CLARIFIER

This activated sludge technology was developed by Advent. It uses a patented (Advent Integrated System (AIS)) baffle type clarifier within the aeration basin and eliminates a separate final clarifier with recycle activated sludge pumping. The technology has cost advantages by eliminating the separate clarifier and recycle pumping. The disadvantage is the recycle of biomass back to the aeration basin cannot be controlled as well as separate recycling pumping. There are several full scale plants operating at refineries and chemical plants including the American Hess plant in New Jersey.

AERATION AND OXYGEN TRANSFER TECHNOLOGIES

Oxygen transfer principles and aeration are covered in detail [1,2] and are not presented in this book. There are numerous types of aeration equipment technologies available for evaluation and selection for industrial wastewater treatment projects. These technologies include the following which are compared in Table 4.9 to oxygen transfer ratings in terms of standard oxygen rating (SOR) expanded in terms of lbs oxygen transferred per wire horsepower under standard conditions of 20°C, zero dissolved oxygen in clean water at sea level:

- Surface aeration (slow and fast speed)
- Coarse bubble diffusers
- Fire bubble diffusers
- Jet aeration

TABLE 4.9. Summary of Aerator Efficiencies.

Type of Aerator	Water Depth (ft)	STE (%)	lb O_2/hp-hr[(1)]
Fine Bubble			
Tubes—spiral roll	15	15 to 20	6.0 to 8.0
Domes—full floor coverage	15	27 to 31	10.8 to 12.4
Coarse Bubble			
Tubes—spiral roll	15	10 to 13	4.0 to 5.2
Spargers—spiral roll	14.5	8.6	3.4
Jet aerators	15	15 to 24	4.4 to 4.8
Static aerators	15	10 to 11	4.0 to 4.4
	30	25 to 30	6.0 to 7.5
Turbine	15	10 to 25	–[(2)]
Surface Aerator			
Low speed	12	–	5.9 to 7.5
High speed	12	–	3.3 to 5.0

- Turbines
- Pure oxygen
- Brush mechanical aerator
- Static aerators

One of the key factors affecting oxygen transfer is the alpha factor α. The alpha in the main transfer rate in wastewater KLA divided by the main transfer rate in clean water. The alpha is influenced by the turbulence in the aeration basin and by the wastewater characterization such as surface active agents which are concentrates at the gas-liquid interface and can both reduce oxygen transfer but also reduce the bubble size which increases oxygen transfer. The alpha factor which can vary from 0.40 to 1.2 should be developed for testing prior to design of the aeration system.

AMMONIA AND NITROGEN REMOVAL TECHNOLOGIES

A number of innovative approaches to biological treatment of the recycle stream for ammonia or total nitrogen removal have been developed in recent years (Jeyanayagam, 2013) [7]. These strategies use the following basic processes either singly or in combination:

- Oxidation of ammonia to nitrite (nitritation)
- Reduction of nitrite to dinitrogen gas (denitritation)

- Oxidation of ammonia to nitrate through nitrite (nitrification)
- Anaerobic oxidation of ammonia to dinitrogen gas (deammonification)

Some of the processes also provide the benefit of supplying a source nitrifying biomass to the main-stream process (bioaugmentation). A brief description of each technology is provided below.

SHARON PROCESS

The SHARON Process is another form of activated sludge process which has been applied to sidestream treatment of centrate from sludge dewatering centrifuges in municipal treatment plants. The centrate has high pH, temperature and ammonia concentrations which are typically recycled back to the main activated sludge process. The centrate wastewater originates from the centrifuge dewatering of sludge from anaerobic digests. The NYC's Wards Island plant utilizes the SHARON Process which was developed at Delft University of Technology in the Netherlands and removes the nitrogen biologically from the centrate at 35°C in a relatively short detention time of 1 to 2 days. A completely mixed reactor without solids recycle and intermittent aeration is used to produce more rapid growth of ammonia oxidizing bacteria versus nitrite oxidizing bacteria to favor production of nitrite versus nitrates. Methanol can be added to a pre-anoxic zone to provide an electron donor for denitrification. The SHARON Process can be operated without the anoxic step to produce a nitrite-rich recycle stream that can be fed to a pre-anoxic zone.

SHARON is one of the first technologies of the current generation of sidestream treatment processes to be developed and, thus, has the longest track record. A schematic of the process is shown in Figure 4.17. Distinguishing features of the process are as follows:

- Elevated recycle stream temperature inhibits nitrite oxidation
- Once-through process—no sludge retention
- Insulated/heated reactor
- Denitritation to dinitrogen gas with methanol addition
- SRT of 1–2 days
- One reactor with cycling of air or two separate rectors

Full-scale SHARON reactors are operating at Zwolle, Beverwijk, and Groningen in the Netherlands. Two more projects have been designed: one for Haag in the Netherlands and the other for the Wards Island WPCP in New York City.

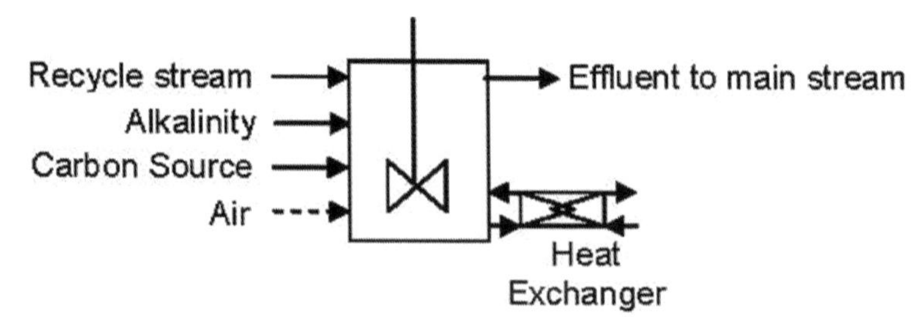

Figure 4.17. SHARON process schematic.

ANAMMOX® PROCESS

The Anammox Process discovered in the mid 1990s utilizes a unique bacteria to convert ammonia to gaseous nitrogen. The bacteria in the Anammox Process are different than the autotrophic nitrifying bacteria in that they cannot use oxygen for ammonia oxidation. Ammonia oxidation with a reduction in nitrite under general conditions has been shown to occur at temperatures above 20°C. This process has been used in SBRs, MBBRs, granular sludge reactors and completely mixed reactors.

The two stage ANAMMOX® process by Paques, uses SHARON process (Hellinga et al, 1998)(8) to convert ammonia to nitrite (partially) followed by the anammox process. The first reactor is optimized to convert ammonia to nitrite by maintaining the proper pH, dissolved oxygen, and taking advantage of the higher growth rate of ammonia oxidizing bacteria at the higher temperatures found in dewatering return water (Van Loosdrecht and Jetten, 1998) (9). Anammox bacteria grows in granules and is retained in the second reactor using upflow clarification.

In the single-stage ANAMMOX® process nitritation and deammonification (anammox) occur in one basin. Process control becomes essential in this arrangement. It is essential to maintain the long SRT for anammox bacteria. Secondly, the process control (pH, DO and ORP) is used to prevent other bacteria competing for nitrite (heterotrophic denitrification and in particular autotrophic NOBs).

Ammonia and nitrite in equimolar concentrations can be oxidized to dinitrogen gas anerobically by bacteria in phylum planctomycetes. An anammox reactor can be coupled with a SHARON reactor to provide feed with the required ammonia and nitrite proportions. A schematic of the process is shown in Figure 4.18. Distinguishing features of the process are as follows:

- Autotrophic—no external carbon source required
- Anaerobic—no aeration required
- Heated/insulated reactor (T = 30–35°C)
- Sludge retention (sequencing batch reactor)

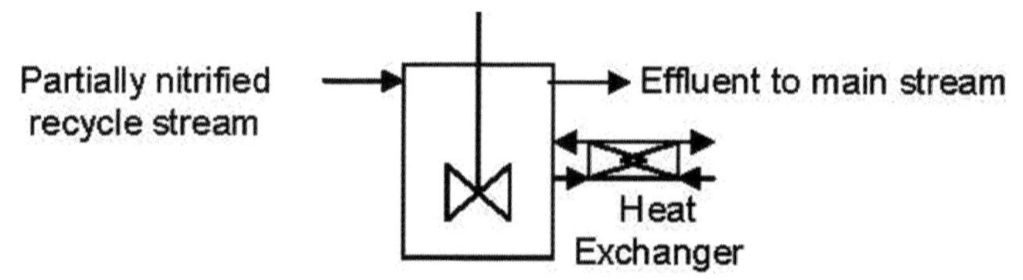

Figure 4.18. Anammox process schematic.

- SRT of 30–50 days
- Requires pre-treatment of feed to equimolar concentration of ammonia and nitrite

A full-scale anammox reactor coupled to a SHARON process is operating in the Netherlands.

DEMON (DeAmmonification)

DEMON incorporates partial aerobic oxidation of ammonia to nitrite and anaerobic oxidation of ammonia and nitrite to dinitrogen gas by the anammox reaction in a single-sludge process. A schematic of this process is shown in Figure 4.19. Distinguishing features of the process are as follows:

- Intermittent aeration to control nitrite oxidation
- Sludge retention (sequencing batch reactor)
- Suspended growth
- Autotrophic denitritation—no external carbon required

DEMON has been demonstrated full-scale at Strass, Austria.

AT-3

This process utilizes a slip-stream of RAS biomass to oxidize ammonia in

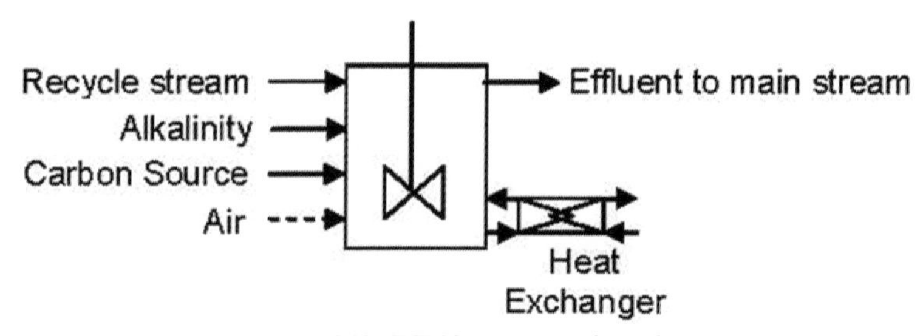

Figure 4.19. DEMON process schematic.

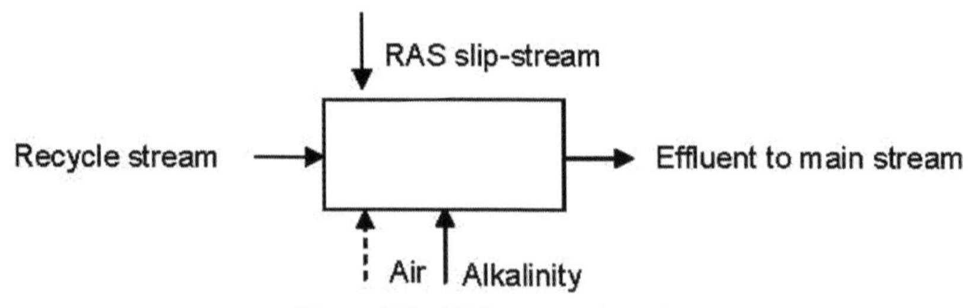

Figure 4.20. AT-3 process schematic.

the recycle stream to nitrite. A schematic of the process is shown in Figure 4.20. Distinguishing features of the process are as follows:

- Free ammonia and nitrous acid toxicity inhibits nitrite oxidation
- Once-through process—no sludge retention
- Open tank
- Potential for bioaugmentation
- SRT of 1–3 days

MAUREEN (MAINSTREAM AUTOTROPHIC RECYCLE ENABLING ENHANCED N REMOVAL)

MAUREEN is a modified version of AT-3 that incorporates an anoxic compartment for denitritation and solids retention through recycle of effluent biomass from the end of the tank to the beginning of the tank. Methanol is required as a source of organic carbon. A schematic of the process is shown in Figure 4.21.

STRASS

The STRASS process is a sequencing batch reactor (SBR) nitritation pro-

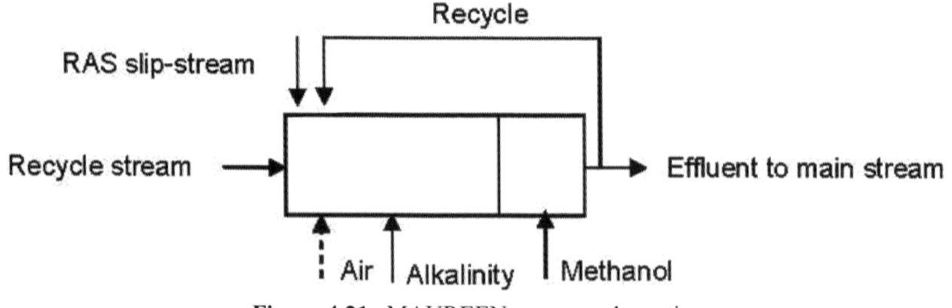

Figure 4.21. MAUREEN process schematic.

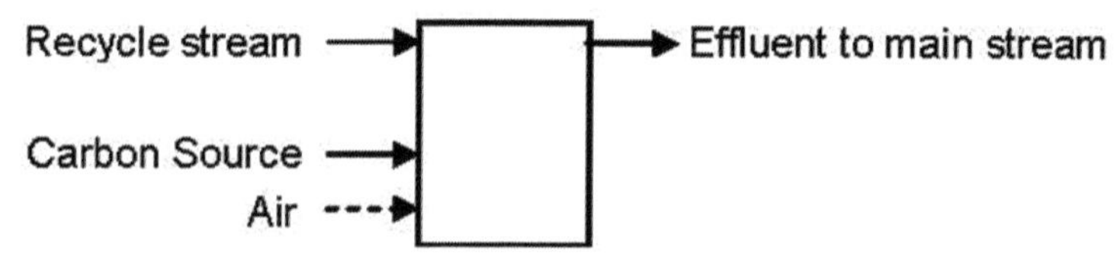

Figure 4.22. SAGB process schematic.

cess that utilizes pH control to inhibit nitrite oxidation. Distinguishing features of the STRASS process are as follows:

- pH control, low dissolved oxygen to inhibit nitrite oxidation
- Sludge retention (sequencing batch reactor)
- High SRT
- External carbon source (primary sludge)

The STRASS process is named for Strass, Austria where the process was first utilized full-scale. Salzberg, Austria is the other full-scale operation. The Strass operation has been converted to the DEMON process. The process has been piloted by the Alexandria, Virginia Sanitation Authority in the U.S.

SUBMERGED ATTACHED GROWTH BIOREACTOR (SAGB)

SAGB is a single-unit, single-zone submerged attached growth bioreactor with intermittent aeration and external carbon addition. A schematic of the process is shown in Figure 4.22.

Operating experience is limited to a single pilot-scale evaluation at the Deer Island WWTP in Massachusetts, USA.

BABE (BIOAUGMENTATION BATCH ENHANCED)

BABE, shown in Figure 4.23, was developed in the Netherlands by DHV in association with Delft University of Technology. Recycle stream from sludge

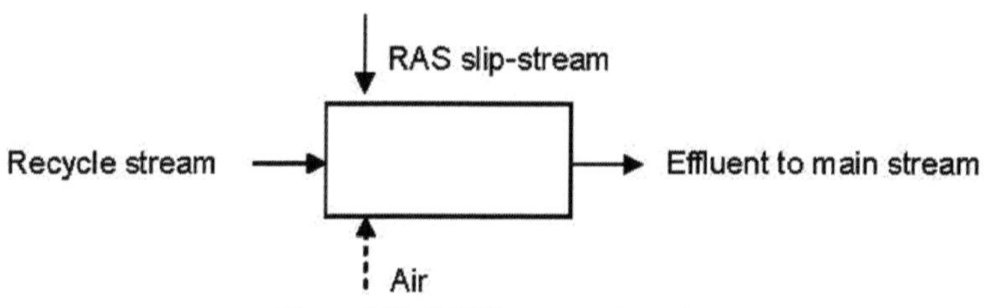

Figure 4.23. BABE process schematic.

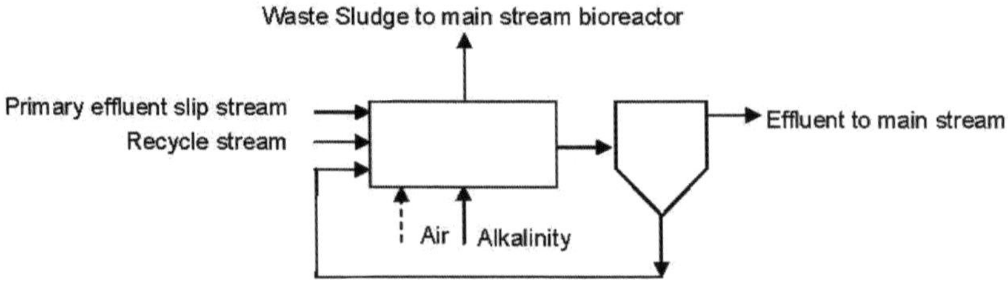

Figure 4.24. In-Nitri process schematic.

digestion is mixed with return activated sludge and is reacted tin a batch mode. Distinguishing features of the process are as follows:

- Sludge retention (sequencing batch reactor)
- Potential for bioaugmentation
- SET < 1 day

BABE was recently demonstrated full-scale at Garmerwolde WWTP in the Netherlands.

IN-NITRI

In-Nitri is a side-stream activated sludge system. Recycle stream ammonia is oxidized to nitrate. Sludge is wasted to main-stream treatment for a bioaugmentation effect. A slip-stream of primary effluent is blended in to acclimate side-stream bacteria to the main-stream. A schematic of this process is shown in Figure 4.24. Distinguishing features of the process are as follows:

- Sludge retention (clarifier)
- Potential for bioaugmentation

There are no full-scale demonstrations of this technology.

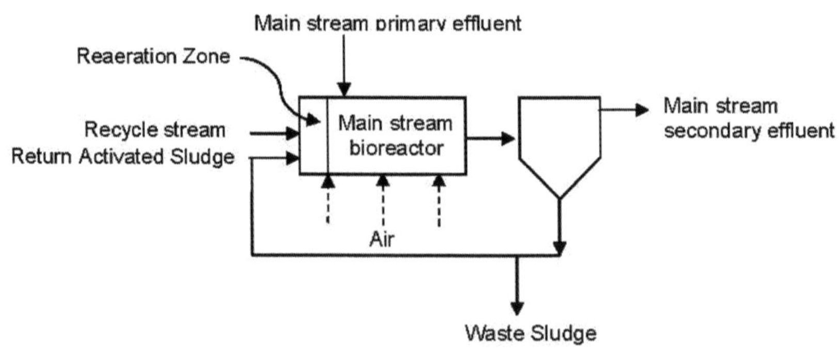

Figure 4.25. BAR process schematic.

BAR (BIOAUGMENTATION REGENERATION/REAERATION)

The BAR process incorporates the discharge of recycle stream into the sludge reaeration zone of the main-stream bioreactor. The stream is fully nitrified by endogenous nitrifiers, which are then carried forward to charge the main aeration tank, thereby reducing the SRT required for complete nitrification. A schematic of this process is shown in Figure 4.25. Distinguishing features of the process are as follows:

- Once-through process—no sludge retention
- Potential for bioaugmentation

The process has seen very wide use in the Czech Republic. The technology was also developed independently in for Appleton, Wisconsin, USA.

TREATMENT OF INDUSTRIAL WASTEWATERS IN MUNICIPAL-ACTIVATED SLUDGE PLANTS

Municipal wastewater is unique in that a major portion of the organics is present in suspended or colloidal form. Typically, the BOD in municipal sewage will be 33 percent suspended, 33 percent colloidal, and 33 percent soluble. The BOD of industrial wastewaters, however, is frequently 100 percent soluble. In an activated sludge plant treating municipal wastewater, the suspended organics are rapidly enmeshed in the floes, the colloids are adsorbed on the floes, and a portion of the soluble organics are absorbed by the biofloc. These reactions occur in the first few minutes of wastewater-biomass contact in the aeration basin. By contrast, for readily degradable industrial wastewaters, a portion of the BOD is rapidly sorbed, and the remainder is removed as a function of time and biological solids concentration. Very little sorption occurs in treatment of refractory wastewaters. The kinetics of the activated sludge process of the municipal plant will, therefore, vary depending on the percentage and type of industrial wastewater in the combined wastewater flow to the plant. The effect on the organic removal rate coefficient (K) caused by addition of an industrial wastewater to a domestic wastewater is almost always determined from treatability studies as discussed in Chapter 7.

The percentage of biological solids in the aeration basin will also vary with the amount and nature of the industrial wastewater contribution. For example, domestic wastewater without primary clarification will yield a sludge that is approximately 47 percent biomass at a 3-day SRT. Primary clarification will increase the biomass percentage to about 53 percent. Increasing the sludge age will also increase the biomass percentage as influent volatile suspended solids undergo degradation and synthesis. Similarly, tile addition of soluble industrial wastewater will increase the biomass percentage in the activated sludge. These conditions will impact flocculation and settling characteristics and the secondary clarification requirements.

The following factors must be considered in the process design for activated sludge treatment of combined domestic and industrial wastewaters:

1. *Effect on sludge quality*—Readily degradable wastewaters will stimulate filamentous bulking, while poorly degradable wastewaters will frequently suppress filamentous bulking. Municipal wastewater itself is subject to filamentous bulking under certain conditions. Addition of a readily degradable wastewater will enhance the potential for bulking, and the use of a selector or a plug-flow configuration may be warranted. Depending on the wastewater mixture, bioinhibitory wastewaters may be effectively treated when mixed with municipal wastewaters in a complete mix configuration.
2. *Effect of temperature*—Addition of an industrial wastewater that has a high soluble substrate load will increase the temperature coefficient, θ, of the combined process. This will decrease process efficiency at reduced mixed liquor operating temperatures.
3. *Sludge handling*—An increase in soluble organics will increase the percentage of biological sludge in the waste sludge mixture. This will decrease thickening and dewaterability, decrease cake solids, and increase chemical conditioning requirements. An exception is pulp and paper mill wastewater, in which pulp and fiber serve as an effective sludge conditioner that enhances dewatering rates but increases the mass of solids for ultimate disposal. An evaluation should be made of changes in sludge handling requirements and production rates that may result from the addition of industrial wastewater.
4. *Bioinhibition and aquatic toxicity*—Many industrial wastewaters inhibit the activated sludge process, particularly with respect to nitrification. Similarly, they may be a new source of potential aquatic toxicity if the toxicants "pass through" the treatment works. It is essential that the industrial wastewater be fully evaluated for compatibility with the existing activated sludge process to insure future permit compliance.
5. *Nutrient (N and P) requirements*—Since many industrial wastewaters are nutrient deficient, the BOD:N:P ratio of the combined wastewater should be determined. If the industrial load contribution is not large, the background excess nutrient concentration of the domestic wastewater may provide the required nutrient balance. If the industrial load is not equalized and fed over time, it can create a shock load with nutrient determining and promote a viscous or polysaccharide bulking problem.

Treatment of frac waters in POTWs was done in 2008 to 2009 and then stopped due to availability of other alternatives. Some states, such as New

York and Pennsylvania, have bans on discharge of frac water to POTWs. The USEPA and State regulations have specifically called out disposal of wastewater from shale gas extractions for increased scrutiny and regulations under NPDES. A proposed rule for shale gas is scheduled for 2014. WEF has developed guidelines for POTWs on questions and concerns relative to acceptance of frac water [10]. Chapter 8 discusses treatment of frac waters and applicable technologies.

ANAEROBIC TREATMENT

Anaerobic treatment technology like activated sludge has been used for years on numerous high strength industrial wastewater projects, particularly in the food and beverage industry. The key advantages of the anaerobic treatment technology compared to activated sludge are: lower sludge production, lower power costs, smaller footprint and potential use of the methane gas for energy. There are several anaerobic treatment technology providers and the challenge for industry is how to evaluate and select the best technical process solution. The anaerobic treatment technology mentioned in Chapter 2 under pretreatment is included in this Chapter since it can be used in combination with activated sludge as pre-treatment to reduce the size of the activated sludge system.

For some industrial wastewaters, anaerobic treatment provides significant advantages and cost payback. It has been used at approximately 1600 industrial facilities in a variety of industries [11] including about 900 in the food and beverage industry. Reactor configurations have evolved and improved since the 1960s to better retain biomass, increase COD loading rates and reduce footprint.

- Anaerobic treatment is a process whereby specialized groups of microorganisms, in the absence of oxygen, degrade organic materials to form methane and carbon dioxide
- Methane is available for beneficial reuse or for heating wastewater to improve treatment system performance.
- Sludge production is low (20% of that produced by aerobic processes)
- Costs are lower than aerobic alternatives
- A large number of types and classes of wastewater constituents have been treated by anaerobic processes

Figure 4.26 shows the sequence of steps in anaerobic treatment from liquefaction of insoluble organics to soluble organics by enzymes which are then converted to volatile acids by acid–providing bacteria. The volatile acids are then converted to methane and carbon dioxide plus bacterial cells by methane producing bacteria.

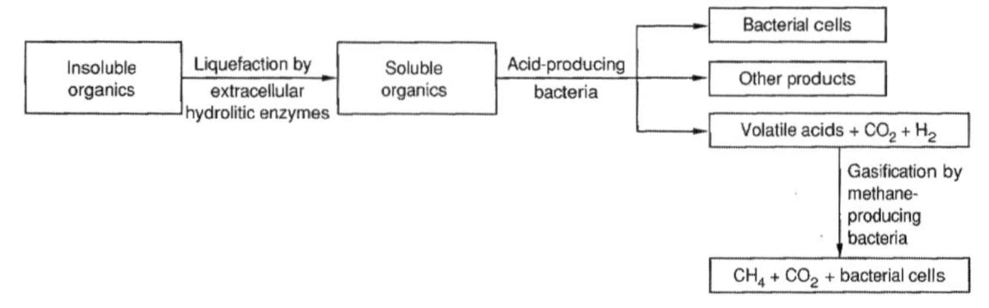

Figure 4.26. Anaerobic treatment sequence.

Table 4.10 shows the history of anaerobic treatment starting in the 1940's with low-rate digestion of municipal sludge. Anaerobic treatment was first used for industry in the 1960's in a meat packaging and food plant (Young, 2003)[12].

Table 4.11 provides a comparison of advantages and disadvantages of anaerobic versus aerobic treatment (Speece, 1996)[13].

AVAILABLE TECHNOLOGIES

Figure 4.27 shows the anaerobic reactor which is typically part of an overall treatment system which includes: equalization; pre-conditioning; heat exchange; gas collection and treatment; and post treatment (Zurawick, 2007) [14].

There are several treatment system configurations. Each involves unique design features to immobilize the biomass in the system. Technologies

TABLE 4.10. History of Anaerobic Treatment.

1940's	Low rate anaerobic municipal sludge digesters
1950–60's	Two stage and heated digesters
1960–70's	Anaerobic contact process used in meat packing and food industry wastewater.
1965–85	Development of Egg-Shaped Digesters for municipal sludges
1966–68	Fixed film reactor lab studies
1972	First full-scale anaerobic filter with pall rings (Celanese)
1980–85	Hybrid anaerobic filter (ADI) • Down flow filter (Bacardi) • Upflow anaerobic sludge blanket (UASB) • (Biothane & Paques) • Fluidized Bed
1990s	Expanded Granular Sludge Bed (EGSB) (Biothane)
1990s–2000	Pulsed sand bed system (Ecovation)

TABLE 4.11. History of Anaerobic Treatment.

Advantages • Volumetric organic loading rates 5–10 times higher than for aerobic processes • Biomass synthesis rates of only 5–20% of those for aerobic processes • Anaerobic biomass preserved for months or years without serious deterioration in activity • No aeration energy requirements for anaerobic processes vs. 500–2000 kw-hr/1000 kg COD for aerobic processes • Methane production of 12,000,000 BTU/1000 kg COD destroyed
Disadvantages • Longer start-up period than aerobic systems • Potential odor with high sulfate wastewaters • Many need heating to operate at 35°C • No nitrification • Effluent quality not suitable for direct discharge • Greater toxicity than aerobic systems for some compounds (e.g., aliphatics)

have improved over the years to retain biomass in the reactor and increase hydraulic and organic loading rates. Some of these technologies can be designed for 20 to 25 kilograms COD per cubic meter reactor volume per day. Figure 4.28 shows several technology or reactor configurations (Speece, 1996)[16].

Case Study No. 1—Anaerobic and Aerobic System Beverage Industry

This project involved the expansion of an existing activated sludge plant to handle an increase in wastewater flow from 40,000 to 100,000 gpd with a COD loading increase of 5 times from 3,700 to 18,500 pounds per day. The COD concentrations of the wastewater averaged approximately 22,000 mg/L.

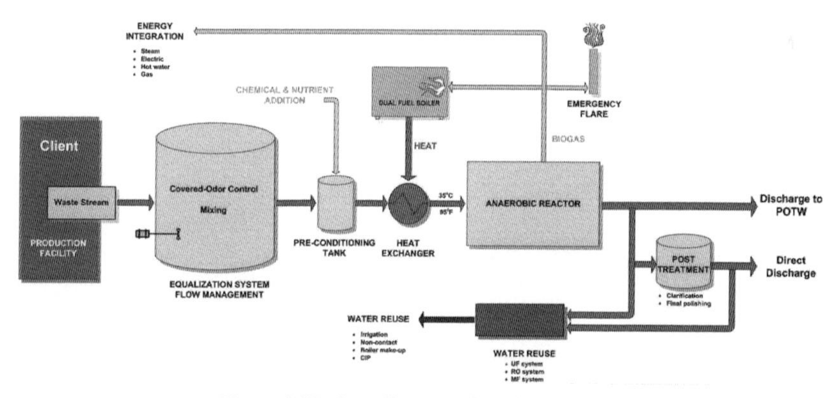

Figure 4.27. Overall process flow diagram [17].

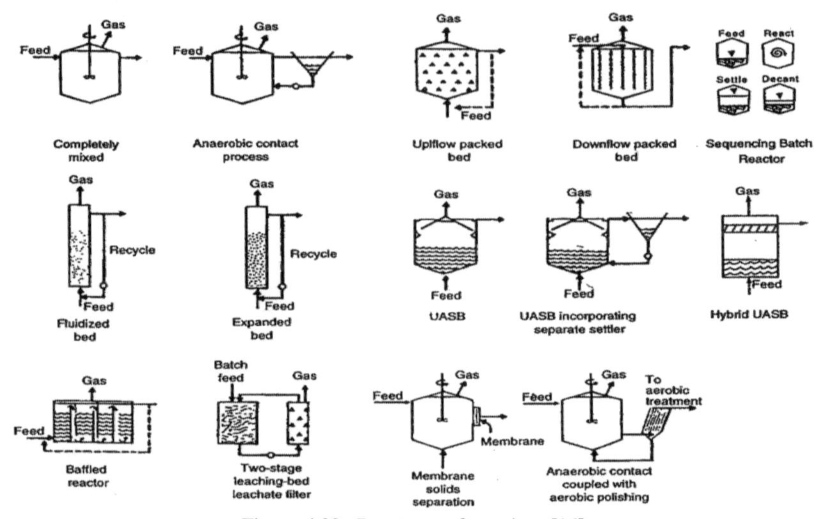

Figure 4.28. Reactor configurations [16].

There was very limited space available, a stringent color limit and the existing facility was over 25 years old. Several alternatives were evaluated to expand the aerobic biological treatment as well as adding anaerobic treatment ahead of the aerobic treatment. The anaerobic treatment option had a higher capital cost versus the aerobic treatment upgrade but a payback of 1 year to 2 years based on lower annual O&M costs primarily due to lower energy and sludge treatment and disposal costs. An anaerobic UASB followed by MBR was recommended as shown in Figure 4.29.

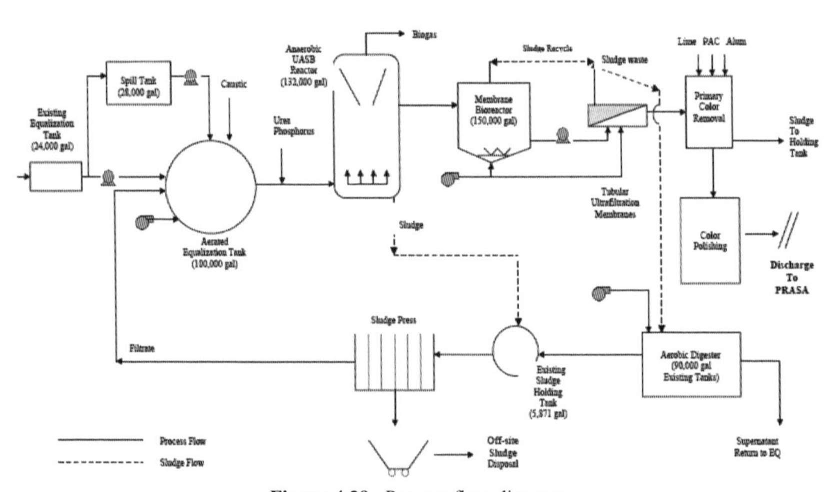

Figure 4.29. Process flow diagram.

Anaerobic MBRs have been used in the dairy, salad dressing and food waste to energy projects as shown in Chapter 4 and as part of water reuse ZLD system in Chapter 7.

CO-DIGESTION AT POTW

Co-digestion, which is becoming more popular, is the addition and treatment of high strength organic liquid industrial wastewater (e.g., winery, dairy, etc.) or fats, oils and grease (FOG) to anaerobic digestion at municipal wastewater treatment plants or POTWs. The addition of organic waste is used to increase the biogas generation for not just heating the digestor but to capture and treat to generate electricity in combined heat and power (CHP) reciprocating engines, micro-turbines and other alternatives. A POTW in upstate New York treats high strength dairy wastewater (COD 60,000 mg/L) along with biosolids in two anaerobic digesters. This plant uses the biogas in a CHP system to produce electricity and achieve essentially net-zero with no need for electricity purchase from the grid.

CO-TREATMENT AT POTW

Other POTWs are now taking these high strength wastewaters directly into the activated sludge process. Two of these plants are experiencing filamentous and viscous bulking which they are fixing with chlorination, polymer and load equalization.

High-strength, readily biodegradable wastewaters such as those from the food processing and brewing industries can be cost-effectively treated first in an anaerobic process, which were discussed in Chapter 3. The anaerobic effluent, however, will not meet typical discharge criteria, and an aerobic process (usually activated sludge) must follow the anaerobic process. The BOD in the anaerobic effluent will consist primarily of acetic acid and other short-chain volatile fatty acids. Since these acids are readily degradable, aerobic process design must consider both effluent quality and sludge quality control. Performance data for a number of industrial wastewaters using low-rate anaerobic treatment followed by activated sludge are shown in Table 4.12.

REFERENCES

1. *Wastewater Engineering Treatment and Reuse.* International Edition, Metcalf and Eddy. Fourth Edition, 2004.
2. Eckenfelder, Jr., W. Wesley, Davis L. Ford and Andrew J. Englande, Jr. 2008. *Industrial Water Quality.* Fourth Edition. McGraw Hill.

TABLE 4.12. Anaerobic-Aerobic Treatment Performance of Industrial Wastewaters [2].[a]

Wastewater	Raw Wastewater			Anaerobic Effluent			Aerobic Effluent		
	COD (mg/L)	BOD (mg/L)	TSS (mg/L)	COD (mg/L)	BOD (mg/L)	TSS (mg/L)	COD (mg/L)	BOD (mg/L)	TSS (mg/L)
Potato processing	4,263	2,664	1,888	144	32	70	88	28	31
Yeast; cane molasses	13,260	6,630	1,086	4,420	600	883	1,856	110	265
Corn processing and municipal	5,780	–	–	1,210	–	136	82	17	28
Alcohol stillage – 5	80,000	18,000	–	–	–	–	40,000	400	–
Alcohol stillage – 6	700	150	–	–	–	–	160	28	–
Molasses stillage	65,000	25,000	5,000	15,000	1,250	500	13,000	200	–
Potato processing	8,356	5,300	5,250	1,113	486	708	250	40	100
Olive processing	13,395	5,550	289	2,332	786	212	884	123	196
Pharmaceutical	9,200	4,000	2,400	3,300	850	350	350	20	30
Potato processing	12,489	5,978	9,993	4,692	1,573	2,200	471	59	149

[a]Courtesy of ADI Inc.

3. Wujcik, Walter J. and Alan F. Rozich. "Design and Start-up of an Advanced Thermophilic Treatment System for High Strength Wastewater from a Pharmaceutical Plant."
4. Briddle, T.R., D.C. Coimenhage and J. Stelzig. 1979. "Operation of a Full-Scale Nitrification-Denitrification Industrial Waste Treatment Plant." *J. Water Pollution Control Federation, 51*:1, 127.
5. Kang, C.W., *et al.* 2008. 'Bridging the Gap Between Membrane Bio-Reactor (MBR) Pilot and Plant Studies, *Journal of Membrane Science,* 325."
6. Sparks, Michael and Dan Turner. 2011. "Next Generation External Membrane Technology—Energy Advantages Using Air Lift MBR Process." WEFTEC 2011.
7. Jeyanayagam, Sam. 2009. "What are Strategies Available for Managing Streams from Sludge Operations." WERF.
8. Hellinga, C., A.A.J.C. Schellen, J.W. Mulder, M.C.M. van Loosdrecht and J.J. Heignen. 2001. "The SHARON Process: An Innovative Method for Nitrogen Removal from Ammonium-rich Wastewater." *Water Sci. Technol. 43*(11) 127–134.
9. Van Loosdrecht, M.C.M. and M.S.M. Jetten. 1998. "Microbiological Conversions in Nitrogen Removal." *Water Sci. Technol. 38*(1) 1–7.
10. WEF Fact Sheet. 2013. "Considerations for Accepting Fracking Wastewater at Water Resource Recovery Facilities."
11. Kleerebezan, R. and H. Macore. April 2003. Treating Industrial Wastewater: Anaerobic Digestion Comes of Age." *Chemical Engineering.*
12. Young, James C. April 2003. "Fundamentals and Design of Advanced Anaerobic Process for Industrial Wastewater Treatment." Workshop B, 9th Annual Industrial Wastes Technical and Regulatory Conference, San Antonio, Texas.
13. Speece, Richard. 1996. "Anaerobic Biotechnology for Industrial Wastewater." Vanderbilt University.
14. Zurawick, Walter. 2007. Personal Communication and Presentation.

CHAPTER 5

Nitrogen Removal—Nitrification and Denitrification

THE activated sludge process can be designed and operated to remove ammonia and nitrate-nitrogen via nitrification and denitrification processes, respectively. The processes of biological nitrification is shown in Figure 5.1 and the nitrogen transformation for nitrification and denitrification are illustrated in Figure 5.2. The nitrification process consumes oxygen and alkalinity, and a portion of the nitrogen is directed to biomass synthesis (waste-activated sludge). The denitrification process occurs under anoxic conditions, consumes BOD, and produces alkalinity and new cells.

NITRIFICATION

The reactions which occur during biological nitrification are summarized below:

$$\text{Organic Nitrogen} \xrightarrow{\text{Biohydrolysis}} NH_4^+$$

$$2NH_4^+ + 3O_2 \xrightarrow{\text{Nitrosomonas}} 2NO_2^- + 4H_2O + 4H^+ \text{ New Cells}$$

$$2NO_2^- + O_2 \xrightarrow{\text{Nitrobacter}} 2NO_3^- + \text{New Cells}$$

For 1 g NH_3-N oxidized to NO_3-N,

- 4.33 g of O_2 are consumed.
- 7.15 g of alkalinity (as $CaCO_3$) are destroyed.
- 0.15 g of new cells are formed.
- 0.08 g of inorganic carbon are consumed.

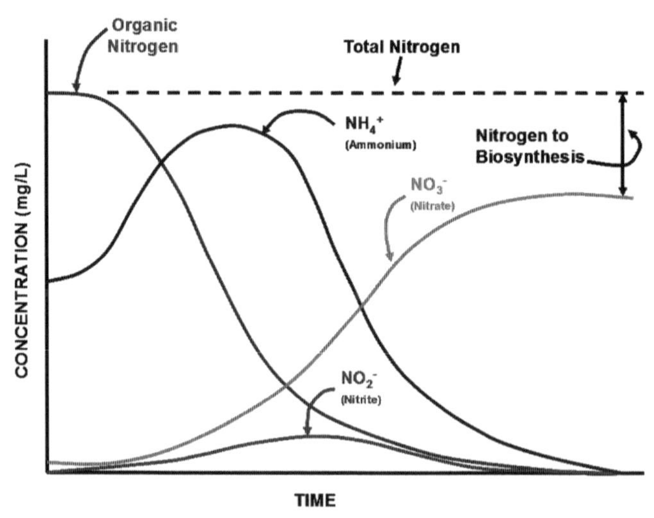

Figure 5.1. Biological nitrogen conversion process.

The biohydrolysis reaction is mediated by a wide range of heterotrophic organisms and seldom limits the rate of nitrogen oxidation. The oxidation of ammonia to nitrate is a sequential reaction. It is carried out under strictly aerobic conditions by only a few species of chemoautotrophic organisms (*Nitrosomonas* and *Nitrobacter*), which derive their energy from the oxidation reaction and their carbon source from alkalinity. As such, they are more sensi-

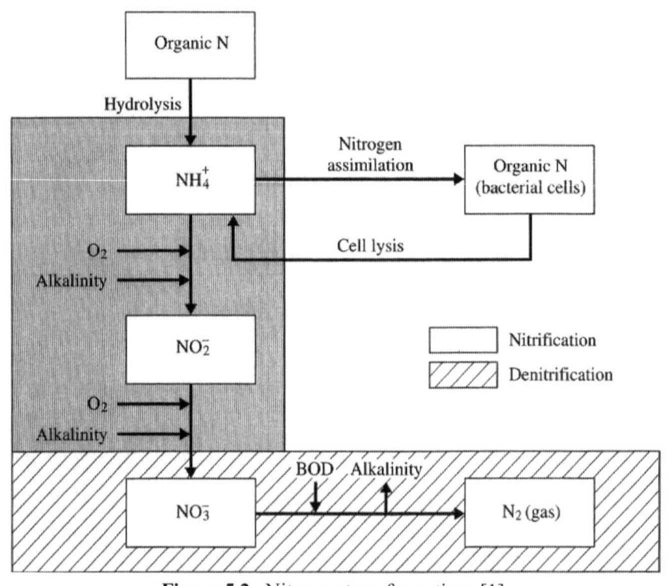

Figure 5.2. Nitrogen transformations [1].

tive to the mixed liquor conditions of pH, temperature, toxics, etc., and grow more slowly than the heterotrophs that consume BOD. When no inhibition exists, the specific growth rate for *Nitrobacter* is higher than the growth rate of *Nitrosomonas*. Hence, there is no accumulation of nitrite in the process, and the growth rate of *Nitrosomonas* will control the overall nitrification reaction.

Nitrification Kinetics

In order to maintain a population of nitrifying organisms in a mixed culture of activated sludge, the minimum aerobic sludge age $(\theta_c)_{min}$ must exceed the reciprocal of the nitrifiers' net specific growth rate:

$$(\theta_c)_{\min} \geq \frac{1}{\mu_{N_T} - b_{N_T}} \tag{5.1}$$

where,

μ_{N_T} = nitrifiers' specific growth rate, days^{-1}
b_{N_T} = nitrifiers' endogenous decay rate, g ΔVSS_N/g VSS_N • days
VSS_N = nitrifier mass expressed as VSS

A safety factor of 1.5 to 2.5 should be applied to the $(\theta_c)_{min}$ value computed from Equation (5.1) to determine the design sludge age $(\theta_c)_d$.

When treating industrial wastewaters that may inhibit nitrification, the maximum specific growth rate must be experimentally determined from the treatability testing discussed in Chapter 7 and shown in Equation (5.2):

$$\mu_{N_T} = a_N q_{N_T} \tag{5.2}$$

where,

q_{N_T} = specific rate of nitrification, g NH_3-N/g VSS_N • day
a_N = gross cell yield coefficient, g VSS_N/g NH_3-N removed

Values of 0.07 to 0.15 g/g have been reported for a_N.

The value of μ_{N_T}, determined by Equation (5.2), is used to determine $(\theta_c)_d$ after adjusting for temperature and applying an appropriate safety factor. The required $X_v t$ is then computed from the relationship:

$$(\theta_c)_d = \frac{X_v t}{a_H S_r + b X_d X_v t} \tag{5.3}$$

For a new plant design, X_v, is selected, and the required aerobic detention time is calculated. For an existing plant, t is fixed, and the required X_v is calculated.

The fraction of nitrifiers in a mixed culture (f_N) is calculated by Equation 5.4:

$$f_N = \frac{a_N N_{OX}}{a_H S_r + a_N N_{OX}} \tag{5.4}$$

in which:

$$N_{OX} = TKN_o - N_{syn} - STKN_e$$

where,

N_{OX} = nitrogen available to be nitrified (mg/L)
N_{syn} = nitrogen directed to cell synthesis ≅ 0.04 S_r (COD basis)
TKN_o = influent TKN, mg N/L
$STKN_e$ = effluent soluble TKN, mg N/L
a_H = sludge yield coefficient for hetrotrophs, mg VSS/mg
a_N = sludge yield coefficient for nitrifiers, mg VSS_N/mg

Nitrification Rate

The rate of nitrification is the same as the ammonia removal rate and can also be calculated as follows:

$$N_r = 1.82 \frac{1}{1 + 0.033\theta_c} \cdot \frac{N_e}{0.4 + N_e} \cdot \frac{DO}{K_o + DO} 1.068^{(T-20)} \tag{5.5}$$

where,

N_r = Nitrification rate, mg NH_3 removed/mg MLVSS-day
θ_c = Sludge age, days
N_e = Effluent ammonia concentration
DO = Dissolved oxygen, mg/L
T = Temperature, °C

The key factors effecting the nitrification rate are:

- Sludge age and *F/M*
- Effluent ammonia concentration
- Dissolved oxygen (DO)
- pH
- Temperature
- Inhibition

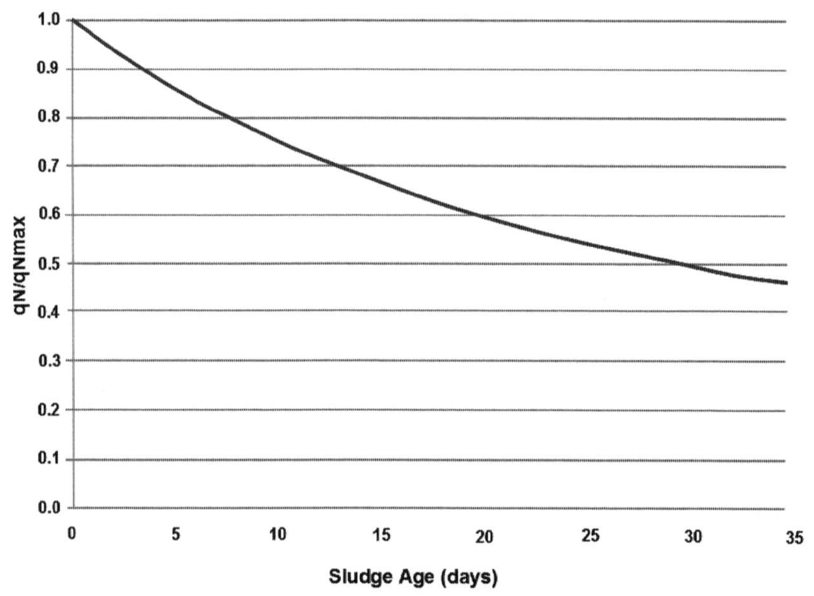

Figure 5.3. The effect of sludge age on the nitrification rate.

Increasing the sludge age oxidizes more of the biomass through endogeneous respiration thereby decreasing the active fraction of nitrifiers in the mixed liquor and the nitrification rate as shown in Figure 5.3.

The effect of effluent ammonia on the nitrification rate is shown in Figure 5.4. Pilot plant rates for a chemical industry wastewater shown in Figure 5.5

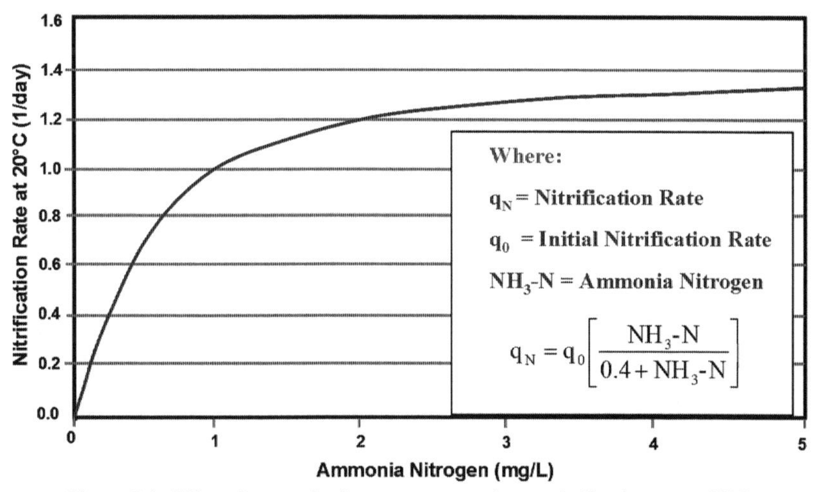

Figure 5.4. Effect of ammonia nitrogen concentration on nitrification rate at 20°C.

shows the zero-order rate at ammonia concentration above 5 mg/L. These data were developed from the second stage of a two-stage activated sludge system.

Figure 5.6 shows the effects of both dissolved oxygen and pH on the nitrification rate. The influence of mixed liquor dissolved oxygen on the nitrification rate has been somewhat controversial, partly because the bulk liquid concentration is not the same as the concentration within the floc where the oxygen is being consumed. Increased bulk liquid dissolved oxygen concentrations will increase the penetration of oxygen into the floc, thereby increasing the rate

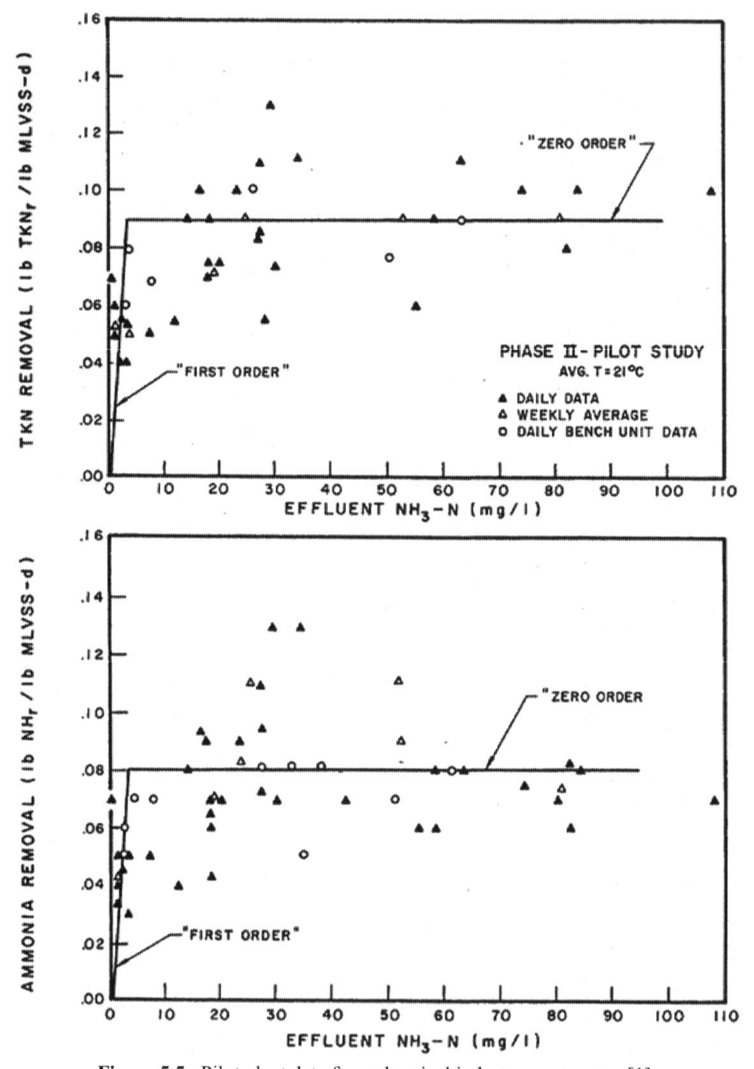

Figure 5.5. Pilot plant data for a chemical industry wastewater [1].

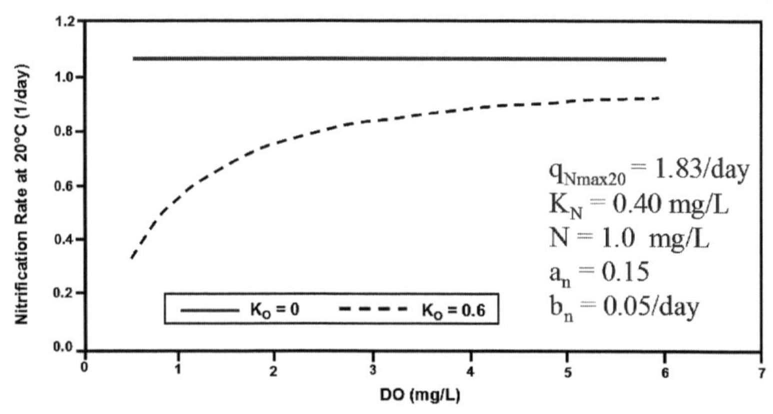

Figure 5.6. The effect of dissolved oxygen on nitrification rate.

of nitrification. At a decreased SRT and higher *F/M*, the oxygen utilization rate due to carbon oxidation increases, thereby decreasing the penetration of oxygen. Conversely, at a high SRT and lower *F/M*, the low oxygen utilization rate permits higher oxygen levels within the floc, and consequently, higher nitrification rates occur. Therefore, to maintain the maximum nitrification rate, the bulk mixed liquor dissolved oxygen concentration must be increased as the SRT is decreased. This is reflected in the coefficient K_o. The effect of dissolved oxygen on nitrification rates is given by Figure 5.6.

Figure 5.7 shows the effect of pH on the nitrification rate and shows pH range of 7.0 to 7.8 as the most favorable range for maximum nitrification.

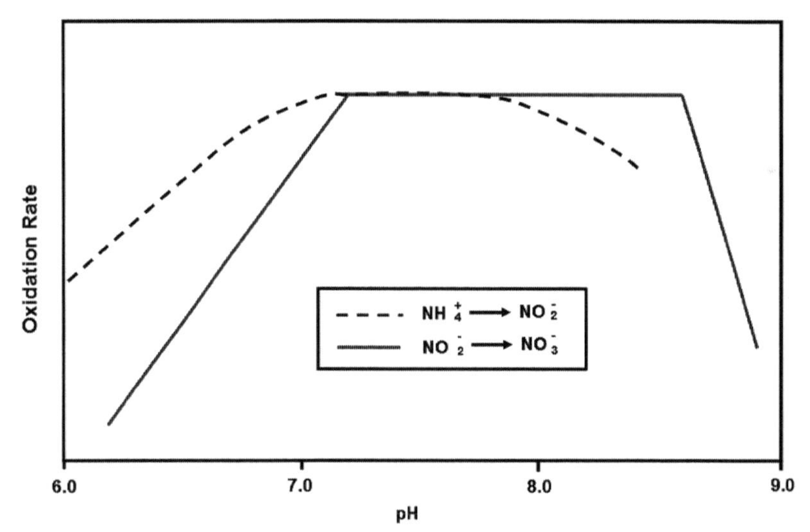

Figure 5.7. The effect of pH on nitrification rate.

INHIBITION

Industrial wastewaters may contain organics or inorganics which inhibit, or even toxic, to the nitrifier organisms. Nitrification results for treatment of an organic chemicals wastewater showed that a minimum aerobic SRT of 25 days was required to obtain complete nitrification at 22 to 24°C [1]. The minimum SRT required for complete nitrification of municipal wastewater at these temperatures is approximately 4 days. An SRT of 55 to 60 days was required for complete nitrification of this wastewater at 10°C versus approximately 12 days for municipal wastewater. It was also shown that, at a mixed liquor temperature of 10°C, the nitrifiers were less tolerant to variations in influent composition and temperature than were the heterotrophic organisms responsible for BOD removal and denitrification. Similar results were obtained for a wastewater from a coke plant in which the nitrification rate was approximately one order of magnitude less than that for municipal wastewater, as shown in Figure 5.8.

The effect of temperature on the minimum sludge age required for nitrification is shown in Figure 5.9. The relationship between nitrification rate and mixed liquor temperature as a temperature correction coefficient, θ, was 1.13. This value is significantly higher than for typical domestic wastewater, indicating that the nitrification rate was more sensitive to the operating mixed liquor temperature.

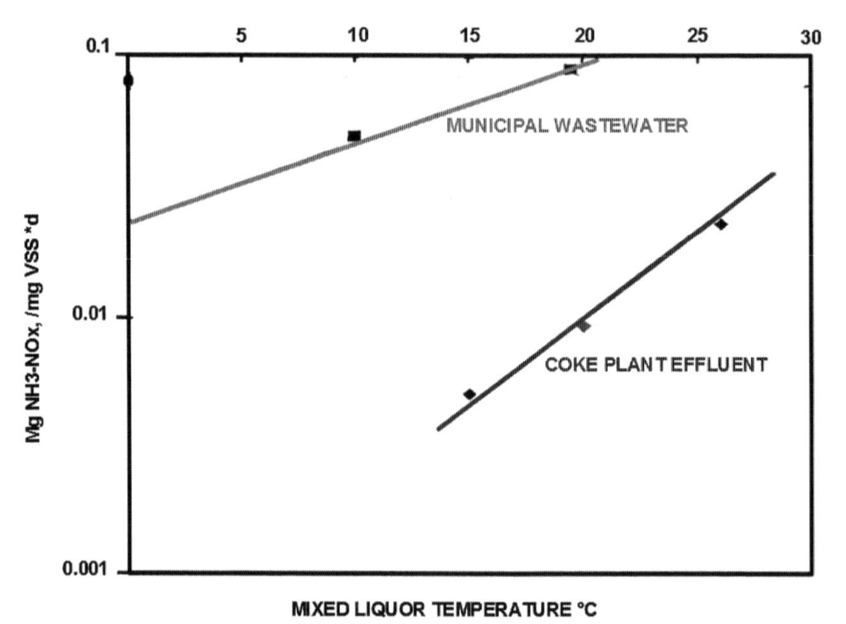

Figure 5.8. Relationship between nitrification rate and temperature for municipal wastewater and a Coke plant effluent.

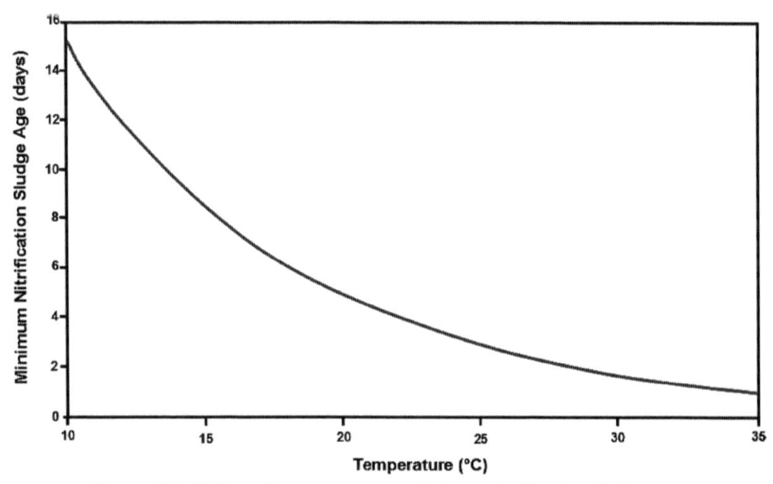

Figure 5.9. Effect of temperature on minimum nitrification sludge age.

In cases where nitrification is significantly reduced or totally inhibited, the application of powdered activated carbon (PAC) to adsorb the toxic agents may enhance nitrification. However, in some cases, excessive quantities of PAC are required to achieve single-stage nitrification, and a second-stage nitrification step can be successfully employed after a first-stage biological process for removal of carbonaceous material and reduction of toxicity. This was the case for the chemical industry example presented previously in Figure 5.5. A plastics additives wastewater with separate-stage nitrification system provided equivalent nitrification efficiency to single-stage treatment using a PAC dose of 1,000 mg/L as shown in Table 5.1.

TABLE 5.1. Single-Stage Versus Separate-Stage Batch Nitrification Testing of Plastic Additives Wastewater [1].

	Nitrogen Concentrations at Indicated Aeration Times					
	t = 0 hrs			*t* = 26 hours		
Treatment Process	**TKN (mg/L)**	**NH_3-N (mg/L)**	**NO_3-N (mg/L)**	**NH_3-N (mg/L)**	**NO_3-N (mg/L)**	**Nitrification Efficiency (%)**
Single Stage @ *F/M* = 0.10 day						
0 mg/L PAC	126	26	1.5	95	0.2	< 1
500 mg/L PAC	134	31	1.2	113	2	1
1,000 mg/L PAC	52	33	1.3	88	72	23
5,000 mg/L PAC	92	34	1.1	75	38	41
Separate Stage @ MLVSS = 1,000 mg/L	105	79	7	45	29	21

Blum and Speece [2] have characterized the toxicity of a variety of organic compounds to nitrification. Unfortunately, many industrial wastewaters contain these and other compounds, which alone or together exert a greater but undetermined inhibition effect on the nitrification process. It is essential, therefore, to determine the specific nitrification rate (q_N) and the $(\theta_c)_{min}$ required to achieve nitrification under actual operating conditions. The value of q_N can be determined using either a batch-activated sludge (BAS) test or fed-batch reactor method which are discussed in Chapter 7.

Un-ionized ammonia (NH_3) inhibits both *Nitrosomonas* and *Nitrobacter*, as shown in Figure 5.10. Since the un-ionized fraction increases with pH, a high pH combined with a high total ammonia concentration will severely inhibit or prevent complete biological nitrification. Since *Nitrosomonas* is less sensitive to ammonia toxicity than *Nitrobacter*, the nitrification process may only be partially complete and result in accumulation of nitrite ion (NO_2^-). This can have severe consequences since (NO_2^-) is strongly toxic to many aquatic organisms whereas (NO_3^-) is not. Ammonia toxicity to activated sludge biomass is rarely a problem when treating municipal wastewaters since the concentration of total ammonia is low and the mixed liquor pH is near neutral. Industrial wastewaters with high ammonia levels and the potential for high pH excur-

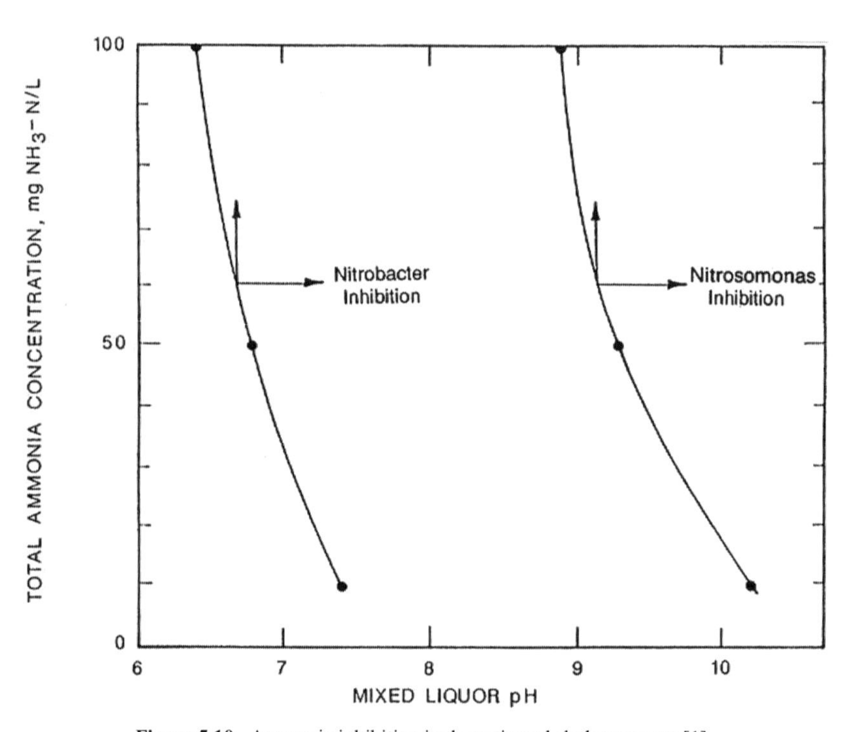

Figure 5.10. Ammonia inhibition in the activated sludge process [1].

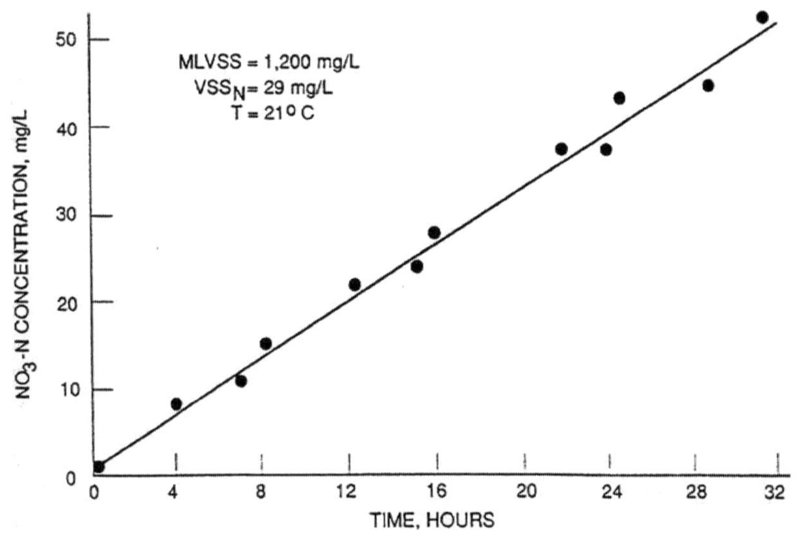

Figure 5.11. Results of batch nitrification test.

sions, however, may cause biotoxicity and loss of the nitrification process. Under these conditions, it is necessary to control the mixed liquor pH to avoid biotoxicity due to an ammonia spill or shock load. In extreme cases, two stages operated at different pH values may be required to separate the *Nitrosomonas* and *Nitrobacter* and allow complete nitrification.

Treatability testing to determine nitrification rates and evaluate inhibitors are discussed in Chapter 7. The results of a batch nitrification test conducted at $T = 21°C$ are shown in Figure 5.11 using a mixed liquor from a municipal wastewater treatment plant. Based on the plant's historical operating data and Equation (5.4), the fraction of nitrifiers (f_N) in the MLVSS was 0.0245 mg VSS_N/mg VSS. Initially, the wastewater had negligible NO_3-N and organic nitrogen and an NH_3-N concentration of 48 mg/L. After 24 hr of aeration, 38 mg/L NO_3-N was produced.

The overall ammonia removal or nitrification rate was calculated as:

$$N_r = \frac{38 \text{ mg/L day}}{1200 \text{ mg VSS/L}} = 0.032 \text{ mg NH}_3\text{-N/mg MLVSS day} \qquad (5.6)$$

The specific nitrification rate was

$$q_N = \frac{N_r}{f_N} \frac{0.032}{0.0245} = 1.3 \text{ mg N/mg VSS}_N\text{-day}$$

The nitrifier specific growth rate for $a_N = 0.10$ mg/mg was 0.13/day. Neglecting temperature effects, the $(\theta_c)_{\min}$ is

$$(\theta_c)_{\min} = \frac{1}{0.13 - 0.05} = 12.5 \text{ days} \tag{5.7}$$

It should be noted that the batch nitrification test used an existing sludge with a known f_N and VSS_N concentration to determine q_N. The actual mixed liquor, established for treatment of this wastewater, will have a different f_N value based on its NH_3-N and BOD concentrations. This nitrifier fraction and the measured q_N should be used to determine the hydraulic retention time required in the aeration basin. The feed batch reactor (FBR) procedure described later in Chapter 7 can also be used to determine the nitrification rate.

Nitrification of High-Strength Wastewaters

Wastewaters, such as some pharmaceutical (see example in this chapter), fertilizer and landfill leachate, containing high ammonia concentrations, as high as 1,000 mg/L, can be treated by biological nitrification. The example at the end of this chapter shows treatment of a pharmaceutical wastewater with both high ammonia and high COD using the MLE activated sludge process.

NITRIFICATION DESIGN PROCEDURE

The following presents a step approach for nitrification system design.

1. Determine the nitrogen to be oxidized:

$$N_{ox} = \text{TKN} - N_{syn} - \text{SON} - \text{NH}_3\text{-}N_{eff}$$

2. Select an operating sludge age and compute $x_{vb}t$:

$$\theta_c = \frac{x_{vb}t}{a_H S_r - b x_d x_{vb} t} \quad \text{or} \quad x_{vb}t = \frac{\theta_c a s_R}{(1 + \theta_c b x_d)}$$

3. Compute the nitrification rate:

$$N_R = 1.82 \frac{1}{(1 + 0.033\theta_c)} \cdot \frac{N_e}{0.4 + N_e} \cdot \frac{DO}{0.45 + DO} \cdot 1.068^{(T-20)} \cdot 1$$

4. Compute the fraction of nitrifiers:

$$f_N = \frac{0.15N_{ox}}{a_H S_r + 0.15N_{ox}}$$

5. Compute the nitrogen oxidized: $N = N_r(f_n)\,(X_{vb})(t)$
6. If the nitrogen oxidized is less than that required, the sludge age must be increased. It if is greater, the sludge age can be decreased.
7. Select the operating sludge age θ_c and compute $x_{vb}t$ and repeat steps 3, 4, 5 and 6. If the nitrogen oxidized equals the nitrogen to be oxidized in No. 1, the sludge age for design is established.
8. Compute the required detention time:

$$t_N = \frac{N_{ox}}{N_r}$$

9. Compute the WAS:

$$\Delta X_{vb} = (a_H S_r - b(1 - f_N)\, X_d f_b t_N) - (a_N N_{ox} - b_N f_N f_b X_d X_V t_N)$$

10. Compute the SRT:

$$\theta_c = \frac{f_b X_v t_N}{\Delta X_{vb}}$$

11. Compute the oxygen requirements for organics removal:

$$\mathrm{O_2/mg/L} = a'S_r + 1\cdot 4bX_d(1 - f_N)(f_b)X_v t_N$$

12. Compute the oxygen requirements for nitrification:

$$\mathrm{O_2/day} = 4.33\cdot N_{ox} + 1.4\ b_N f_N X_d f_b X_v t_N$$

13. Compute the alkalinity consumed:

$$\mathrm{Alk} = 7.15\cdot N_{ox}$$

Design Example 5.1—Nitrification System

Design a nitrification system to produce an effluent NH_3-N of 1 mg/L at 20°C. The influent TKN is 40 mg/L and the BOD_5 210 mg/L.

What must the HRT and SRT be increased to if the temperature is 15°C?

What will the maximum NH_3-N effluent concentration be if the influent NH_3-N increases to 60 mg/L at 20°C?

Conditions:

$DO = 2.0$ m/L

$q_{n(\max)} = 2.3$ mg NH_3-N (mg $VSS_N \cdot$ d)

$a_N = 0.15$

$a_H = 0.6$

$a' = 0.55$

SON = 1 mg/L

$SBOD_c = 10$ mg/L

$X_V = 3{,}000$ mg/L

$b_N = 0.1$ d^{-1}

$b_H = 0.1$ d^{-1}

Solution:

1. Assume a sludge age, $\theta_c = 10$ days.

2. $$X_d = \frac{0.8}{1+0.2\cdot b_H\theta_c} = \frac{0.8}{1+0.2\cdot 0.1\cdot 10} = 0.67$$

3. $$f_{aN} = \frac{1}{1+0.2\cdot b_N\theta_c} = \frac{1}{1+0.2\cdot 0.1\cdot 10} = 0.83$$

4. $$q_N = q_{N\max}\cdot\frac{\text{NH}_3\text{-N}_\text{e}}{0.4+\text{NH}_3\text{-N}_\text{e}}\cdot\frac{DO}{0.45+DO}\cdot f_{aN}$$

5. $$q_N = 2.3\cdot\frac{1}{1.4}\cdot\frac{2}{2.45}\cdot 0.83 = 1.11/d$$

6. The BOD removed is:
   $S_r = S_o - S_e = 210 - 10 = 200$ mg/L

7. The N used for heterotrophic cell synthesis is:
   $N_{\text{Syn}} = 0.08\, a\text{Sr} = 0.08 \cdot 0.6 \cdot 200 = 9.6$ mg/L

8. The N oxidized is:
   $N_{ox} - \text{TKN}_o - \text{NH}_3\text{-Ne} - \text{SON}_e - N_{\text{Syn}}$
   $N_{ox} = 40 - 1 - 1 - 9.6 = 28.4$ mg/L

9. The nitrifier fraction is:

$$f_N = \frac{a_N N_{ox}}{A_N N_{ox} + a_H Sr} = \frac{0.15\cdot 28.4}{0.15\cdot 30 + 0.6\cdot 200} = 0.034$$

10. The nitrification rate is:

$$r_N = q_N f_N X_{vb}$$

$$r_N = 1.11 \cdot 0.034 \cdot 3000 = 115 \text{ mg/(L} \cdot \text{d)}$$

11. The required hydraulic detention time is:

$$t_N = \frac{N_{ox}}{rN} = \frac{28.4}{113} = 0.25 \text{ days}$$

12. The sludge production is:

$$\Delta X_{v_b} = a_H Sr + a_N N_{ox} - [b_H(1-f_N) + b_N f_N] X_d X_{vb} t_N$$

$$\Delta X_{v_b} = 0.6 \cdot 200 + 0.15 \cdot 28.4 - [0.1(1-0.034) + 0.1 \cdot 0.034] \cdot 0.67 \cdot 3000 \cdot 0.25$$

$$\Delta X_{v_b} = 124 - 49 = 75 \text{ mg/L}$$

13. Check θ_c:

$$\theta_c = \frac{X_{v_b} t}{\Delta X_{v_b}} = \frac{3000 \cdot 0.25}{75} = 10 \text{ days}$$

14. Oxygen for Organics removal:

$$\begin{aligned} O_2/\text{mg/L} &= a' S_r + 1.4 b X_d X_{vb}(1-f_N) f_b t_N \\ &= 0.55 \cdot 200 + 1.4 \cdot 0.1 \cdot 0.67 \cdot 3000 \cdot 0.964 \cdot 1 \cdot 0.25 \\ &= 110 + 73 \\ &= 178 \text{ mg/L} \end{aligned}$$

15. Oxygen for nitrification:

$$\begin{aligned} O_2 &= (4.33 \cdot 28.4) + (1.4 \cdot 0.1 \cdot 0.034 \cdot 0.67 \cdot 1 \cdot 3000 \cdot 0.25) \\ &= 123 + 2.4 \\ &= 125.4 \text{ mg/L} \end{aligned}$$

16. Alkalinity consumed:

$$\text{Alk} - 7.15 \cdot 28.4 = 203 \text{ mg/L}$$

At 15°C

$$b_{His} = b_H \; 1.04^{(T-20)} = 0.1 \cdot 1.04^{(15-20)} = 0.082 d^{-1}$$

$$b_{Nis} = 0.082 \; d^{-1}$$

$$q_{Nis} = q_N 1.068^{(T-20)} = 0.82 d^{-1}$$

Assume $\theta_c = 21d$

$$X_{dis} = \frac{0.8}{1 + 0.2 \cdot 0.082 \cdot 21} = 0.60$$

$$f_{vanis} = \frac{1}{1 + 0.2 \cdot 0.082 \cdot 21} = 0.75$$

$$r_{N15} = q_{His} \cdot f_N X_v f_{vanis} = 0.87 \cdot 0.034 \cdot 3000 \cdot 0.74$$

$$r_{H15} = 67 \text{ mg/(L} \cdot \text{d)}$$

$$t_{N15} = \frac{28.4}{67} = 0.42d$$

$$\Delta X_v = 124 - [0.082(1 - 0.036) + 0.082 \cdot 0.036]\ 0.60 \cdot 3000 \cdot 0.42$$

$$\Delta X_v = 124 - 66 = 58 \text{ mg/L}$$

$$\text{Check } \theta_c = \frac{3000 \cdot 0.42}{62} = 20 \text{ days}$$

$$\text{Assuming } q_N = q_{N_{\max}} \cdot \frac{DO}{DO + 0.46} = 2.3 \cdot \frac{2}{2.44} = 1.92/d$$

$$r_{N_{\max}} = 1.92 \cdot 0.036 \cdot 3000 \cdot 0.83 = 172 \text{ mg/(L} \cdot \text{d)}$$

$$N_{ox} = r_{H_{\max}} \cdot t_N = 172 \cdot 0.27 = 46 \text{ mg/L}$$

$$\text{NH}_3\text{-N}_e = 60 - 1 - 8 - 46 = 5 \text{ mg/L}$$

Design Example 5.2—Nitrification (Single vs. Two Stage System)

This design example shows the calculation to compare nitrogen removal in a single stage versus two stage sludge system. The nitrogen removal rate is higher in the first stage of a two-stage system at the higher effluent ammonia concentration.

- Oxidizable nitrogen = 40 mg/l
- BOD = 200 mg/L
- a = 0.55 (heterotrophic yield)
- MLVSS = 3,000 mg/L
- Sludge Age = 12 days
- Active Biomass = 1/1 + 0.033 times θ_c = 0.716
- Assume max nitrification rate = 1.66
- Stage 1 effluent NH_3-N = 5 mg/L

Stage 1 at 5 mg/L NH_3-N in Effluent

1. Oxidizable nitrogen = 40 – 5 = 35 mg/L
   Select sludge age = 12 days

2. Nitrification rate:

$$1.66\left[\frac{1}{1+0.033\ (12)}\right]\left[\frac{1}{5/(0.4+5)}\right]=1.10 \text{ mg Nr/mg MLVSS (Nit)-day}$$

3. Nitrifier fraction:

$$f_n = \frac{0.15(35)}{0.55(200)+\ 0.15(35)} = 0.0456 \text{ or } 4.56\%$$

4. Compute detention time:

$$N = N_r(f_n)(X_{vb})(t)$$
$$35 = 1.1(0.0456)(3000)(t)$$
$$t = \frac{35}{1.1(0.0456)(3000)} = 0.233 \text{ days}$$

Stage 2 at 1 mg/L NH_3-N in Effluent

1. Oxidizable nitrogen = 5 – 1 = 4 mg/L
   Sludge age = 12 days
2. Nitrification rate:

$$1.66\left[\frac{1}{1+0.033(12)}\right]\left[\frac{1}{0.4+1}\right] = 0.85 \text{ mg Nr/mg MLVSS(Nit)-day}$$

3. Nitrifier fraction: same as Stage 1

$$f_n = \frac{0.15(35)}{0.55(200)+0.15(35)} = 0.0456 \text{ or } 4.56\%$$

4. Compute detention time:

$$N = N_r(f_n)(X_{vb})(t)$$
$$35 = 0.85(0.0456)(3000)(t)$$
$$t = \frac{35}{0.85(0.0456)(3000)} = 0.034 \text{ days}$$

Single Stage at 1 mg/L NH_3-N in Effluent

1. Oxidizable nitrogen = 40 – 1 = 39 mg/L
   Sludge age = 12 days

2. Nitrification rate:

$$1.66\left[\frac{1}{1+0.033(12)}\right]\left[\frac{1}{0.4+1}\right]=0.85 \text{ mg Nr/mg MLVSS(Nit)-day}$$

3. Nitrifier fraction:

$$f_n=\frac{0.15(39)}{0.55(200)+\ 0.15(39)}=0.050 \text{ or } 5.0\%$$

4. Compute detention time:

$$N=N_r(f_n)(X_{vb})(t)$$
$$35=0.85(0.050)(3000)(t)$$
$$t=\frac{39}{0.85(0.050)(3000)}=0.306 \text{ days}$$

In summary,

1. Total Complete Mix HRT for 2-Stage System
   0.233 + 0.034 = 0.267 days
2. Total Complete HRT for Single Stage System
   0.306 days
3. Additional Tankage needed for a Single Stage System:

$$\frac{0.306-0.267}{0.267}(100)=14.6\%$$

4. The nitrification rate is a function of sludge age, % of nitrifiers in the MLVSS and the effluent NH_3-N_3;
5. Stage 1 effluent ammonia target provides for increased nitrification rate.
6. A two stage system can have an overall detention time and tank size reduction due to a higher rate in first stage; and
7. Treatability studies are needed for industrial wastewater to determine the nitrification rate.

DENITRIFICATION

Some industrial wastewaters such as those from fertilizer, explosive/propellant manufacture, and the synthetic fibers industry contain high concentrations of nitrates, while others, such as pharmaceutical and leachate, generate nitrates by the nitrification process. Since biological denitrification generates one hydroxyl ion while nitrification generates two hydrogen ions, it may be advantageous to couple the nitrification and denitrification processes to provide "internal" buffering capacity. While many organics inhibit biological nitrification, this is not generally true for denitrification. Sutton *et al.* [3] showed that denitrification rates for an organic chemicals plant wastewater were comparable to those observed using nitrified municipal wastewater; however, biological nitrification of the organic chemicals wastewater was severely inhibited. Denitrification uses BOD as a carbon source for synthesis and energy and nitrate as an oxygen source.

$$NO_3^- + BOD \rightarrow N_2 + CO_2 + H_2O + OH^- + \text{New Cells}$$

The denitrification process consumes approximately 3.7 g COD/g NO_3-N reduced and produces 0.45 g VSS and 3.57 g alkalinity per g NO_3-N reduced.

The rate of denitrification (R_{DN}) is zero order to NO_3-N concentrations of approximately 1.0 mg N/L and is determined by Equation (5.5).

$$R_{DN} = \frac{NO_3 - N_o\text{-}NO_3\text{-}N_e}{X_v t} \tag{5.8}$$

where,

R_{DN} = denitrification rate, g NO_3-N/g VSS-day
X_v = nonnitrifier biomass under aeration, mg VSS/L

The denitrification rate is adjusted for mixed liquor temperature and bulk dissolved oxygen (DO) concentration by Equation 5.6. The temperature coefficient (θ_{DN}) varies from 1.07 to 1.20.

$$R_{D_{NT}} = R_{DN\,20°}\theta_{DN}^{(T-20)}(1 - DO) \tag{5.9}$$

Results of denitrification of an organic chemicals plant wastewater are shown in Figure 5.12.

The factors affecting the denitrification rate include: temperature, sludge age, substrate biodegradability and dissolved oxygen. The denitrification rate will depend on the biodegradability of the organics in the wastewater and the concentration of active denitrifying biomass under aeration. This, in turn, is related to both the SRT and *F/M* and the presence of inert solids in the sludge. Figure 5.13 shows the effect of the *F/M* on the rate of denitrification. As the *F/M* increases, the concentration of active biomass and the rate of denitrification increase. Although denitrification can occur under endogenous conditions (low *F/M*) using internal biomass reserves, it is very slow and requires long hydraulic retention times. Since the rate of denitrification is affected by both wastewater characteristics and process design parameters, it is usually neces-

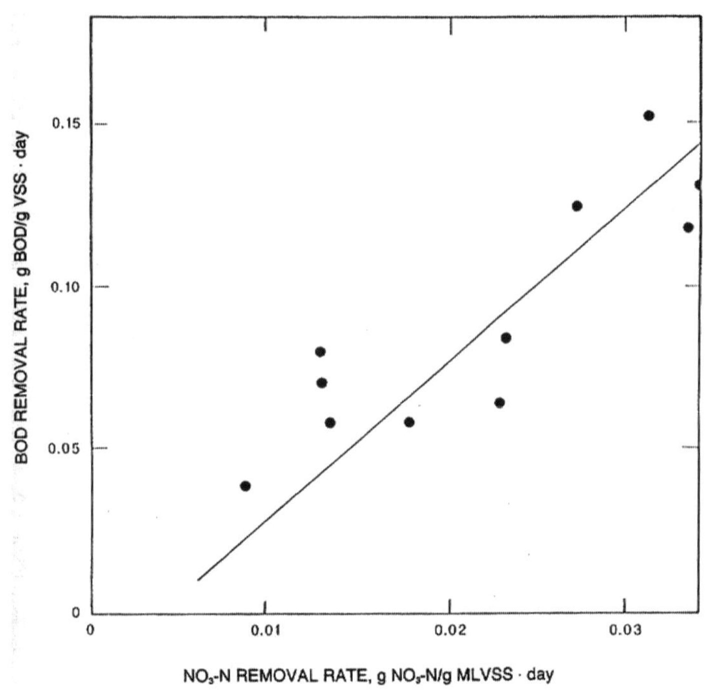

Figure 5.12. Relationship between nitrate reduction and BOD removal for an organic chemicals wastewater [1].

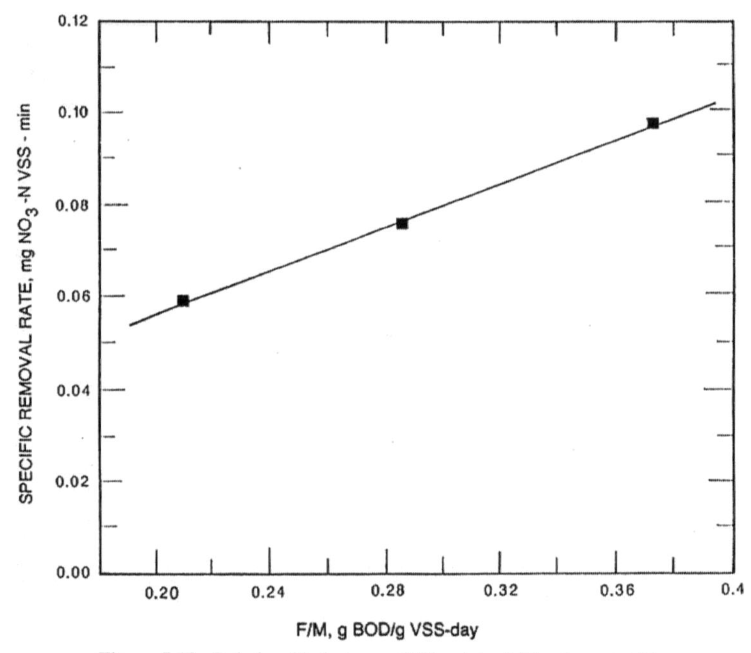

Figure 5.13. Relationship between *F/M* and denitrification rate [1].

sary to determine the rate by experimental means. A batch denitrification test should be conducted in which nitrified effluent containing nitrates, activated sludge and wastewater are mixed under anoxic conditions (ORP ≈ −100 mv) and the residual NO_3-N concentration is determined with time. Depending on the organic composition of the wastewater, one of several removal rate relationships may be obtained.

The denitrification rate for a wastewater can also be estimated from the oxygen uptake rate. In this case, the wastewater-anoxic sludge mixture is aerated and the SOUR determined over time. Correlation of R_{DN} and SOUR indicates that 1.0 mg NO_3-N is equivalent to approximately 3.0 mgO_2, which is in good agreement with the theoretical value of 2.86 mg NO_3-N/mg O_2.

COMBINED NITRIFICATION AND DENITRIFICATION SYSTEMS

A number of alternative treatment systems are available to achieve nitrification and denitrification, in which some form of aerobic-anoxic sequencing is provided. The systems differ in whether they utilize a single sludge or two sludges in separate nitrification and denitrification reactors. The single-sludge system uses one basin and clarifier and the raw wastewater or endogenous reserves as the carbon and energy sources for denitrification.

The two-sludge system uses two basins with separate clarifiers to isolate the

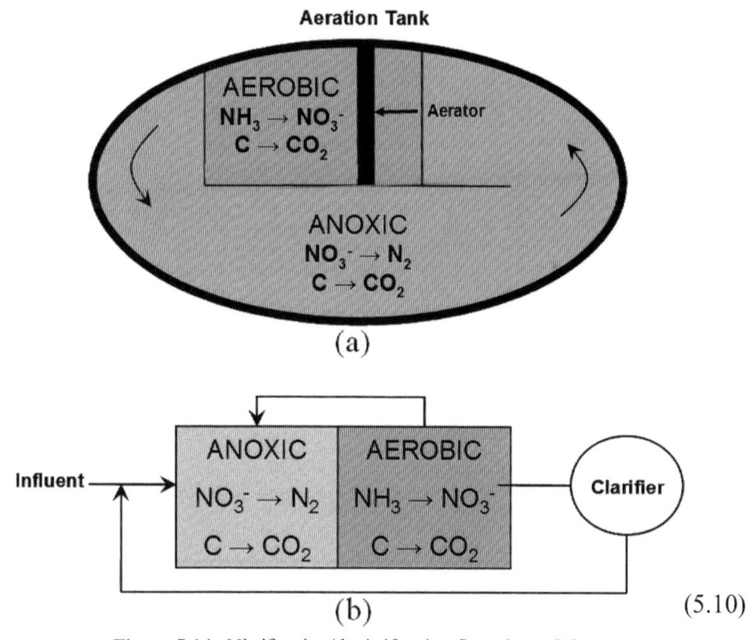

(5.10)

Figure 5.14. Nitrification/denitrification flow sheets [1].

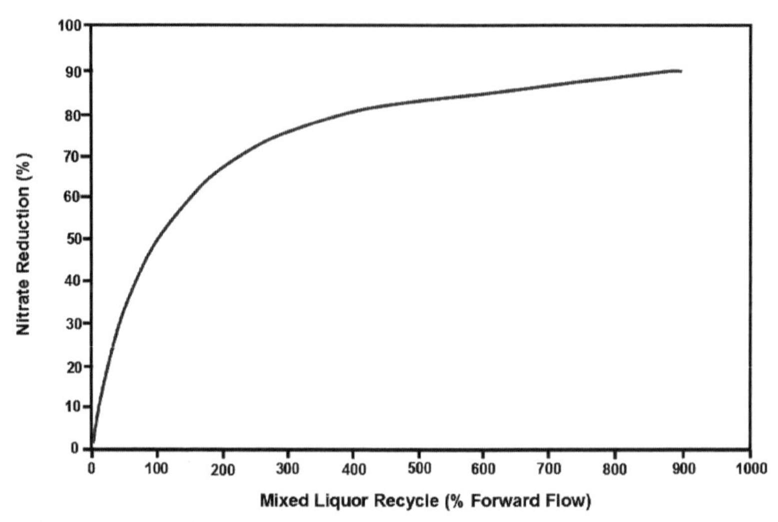

Figure 5.15. Overall nitrate reduction as a function of the internal mixed liquor recycle.

the sludges. A supplemental carbon source such as methanol (CH_3COOH) is provided to the second stage for the carbon and energy source. The simplest configuration of the single-sludge system enables carbonaceous oxidation, nitrification, baffles, mixers and denitrification to occur in a single basin by positioning the return sludge, baffles, mixers and aeration equipment to maintain defined aerobic-anoxic zones in different basin sections. An alternative single-sludge system utilizes a single basin for both aeration and sedimentation by providing intermittent aeration and nonaeration cycles to yield aerobic and anoxic phases of sufficient duration to permit nitrate reduction. Two process flow configurations for single-sludge nitrification and denitrification are shown in Figure 5.14. In the oxidation ditch [Figure 5.14(a)], an aerobic zone exists in the vicinity of the aerator. As the mixed liquor passes away from the aerator, the dissolved oxygen is depleted. Anoxic conditions then exist, and denitrification occurs. This sequence is repeated around the ditch at each aerator.

In the "internal recycle" single-sludge process [Figure 5.14(b)], nitrification occurs under aerobic conditions in the second basin. The second basin may be a separate tank (without intermediate clarification) or a single tank with internal baffles to isolate the aerobic and anoxic zones without short-circuiting. Each of these zones can be plug-flow or CMAS. An internal recycle flow (Q_{Rin}) is employed from the end of the aerobic basin to the inlet of the anoxic basin in which denitrification occurs. The percent of denitrification relative to the total recycle ratio can be calculated by Equation (5.10) and is shown in Figure 5.15.

$$R = \left[\frac{\text{TKN}_o - (\text{NH}_3\text{-N})_{\text{syn}}}{(\text{NO}_3\text{-N}_e) - (\text{NH}_3\text{-N})_e} \right] - 1.0$$

where,

R = total return sludge rate, percent
= $(Q_R + Q_{Rin})/Q_o$
TKN_o = influent TKN, mg/L
$(NH_3\text{-}N)_e$ = effluent NH_3-N, mg/L
$(NO_3\text{-}N)_e$ = effluent NO_3-N, mg/L
$(NH_3\text{-}N)_{syn}$ = nitrogen utilized for biomass synthesis, mg/L

DENITRIFICATION DESIGN PROCEDURE [4]

1. Determine nitrogen to be oxidized
   $TKN - NH_3\text{-}N_{eff} + N_{syn}(0.04BOD_R) + SON$
2. Compute NO_3-N to be denitrified
3. Estimate simultaneous N/DN and recycle NO_3-N
   - Simultaneous N/DN is a function of dissolved oxygen in the aeration basin
   - The NO_3-N denitrified in the anoxic basin is a function of the internal recycle plus the return sludge recycle.
4. The detention time required for denitrification is computed from the relationship:

$$\frac{3(NO_3\text{-}N)}{x_{vb}\, f_a t} = K_{DN}\frac{S_e}{S_o}$$

In which NO_3-N is the concentration of nitrate to be denitrified. Design examples for denitrification are presented by Eckenfelder *et al.* [4].

Design Example 5.3—Upgrade Nitrification/Denitrification Activated Sludge MLE Process to MBR

This example or case study involved an expansion of an existing MLE process treating pharmaceutical wastewater with high COD and TKN concentration of 8,000 and 1,000 mg/L, respectively. The alternatives evaluated included adding additional anoxic and aeration basin volume and final clarification versus conversion to an MBR system. The following questions need to be addressed when comparing the two alternatives:

Questions to ask for MLE Plant Expansion:

1. Is additional aeration basin volume needed to treat the increase in COD and TKN loads?

2. Has the wastewater characterization changed? If so, could the COD and nitrification kinetics change versus the current design? Is treatability testing needed?
3. How much additional aeration basin volume is needed to satisfy nitrate removal versus transfer enough oxygen to maintain dissolved oxygen concentrations in the aeration basin. Which of these controls the aeration basin sizing?
4. What is the anoxic basin volume needed for the projected COD loads and nitrate recycle needed to meet effluent nitrate objectives?
5. Is additional final clarification needed?

Questions to Convert to MBR:

1. Where would membranes be located? In existing final clarifiers, aeration basins or are separate membrane tanks needed to facilitate membrane clearing?
2. What design MLSS concentration should be selected for the existing aeration basins for MBR process?
3. How does MLSS design criteria change for different membrane systems? (i.e., spiral wound versus flat plate)
4. Is the existing aeration basin volume sufficient at design MLSS concentration to treat the nitrogen loads?
5. Can the existing aeration system (i.e., blowers and diffusers), support the design oxygen requirements?
6. What is the maximum oxygen transfer capability of the current aeration system?
7. What are the alternatives for upgrading the aeration system, if needed?
8. What are the oxygen requirements in the separate membrane tank or final clarifiers versus immersed membrane in the aeration basin?
9. What are the differences in cleaning requirements and facilities for each of the various membrane technology providers?
10. What are the cost and non-costs selection criteria for process alternatives selection and the membrane system selection for the MBR alternatives if selected?
11. How does one select a MBR technology provider with so many available?
12. What is the alternative MBR provider's life cycle cost comparisons including membrane replacement?
13. How does one construct the upgrade while keeping the current plant operated?

A process model (i.e., Biowin or GPSX) verification plus process calculations done by hand are recommended to understand the key process design criteria, which drives the unit sizing of the key facilities and equipment and to make the process selection decision. The hand calculations are always a good training tool for the process engineer and operators to better understand how the treatment plant performs as process variables and loadings change. The process model can be used to evaluate the sensitivity of the flows, loads and process design criteria to unit sizing and process performance.

Here is an example of the process design calculations that are needed in this example.

Option 1—Expand Existing Activated Sludge MLE Plant

Future Design Loads

Flow, gpd	500,000
BOD, lbs/day	15,000
BOD, mg/L	3,600
TKN, lbs/day	4,000
TKN, mg/L	960
Design Temperature in Aeration Basin	30°C
Existing Aeration Volume	0.5 MG

Nitrification Removal Rate:

Assume 1.0 lbs TKN/lb MLVSS-day @ 30°C developed from treatability testing
4,000/1.0 = 4,000 lbs MLVSS
$TKN_{ox} = 4,000 - 0.04\ (15,000) = 3,400$

Minimum Aeration Tank Volume Based on Nitrifiers Present:

Assume MLSS	= 5,000 mg/L
Assume MLVSS	= 0.9 MLSS = 4,500 mg/L
Assume % Nitrifers in MLVSS	= 7% (from Biowin model work)
MLVSS Nitrifers	= 0.07 (4,500) = 315 mg/L
Aeration Basin Volume (MG)	= 3,400/(0.07)(5,000)(0.9)(8.34)

Required Volume = 1.29 million gallons (2.6 times existing volume)
Additional aeration volume needed, 1.29 – 0.5 Existing = 0.79 MG

Calculate Aeration Basin Volume to Supply Oxygen and
Satisfy Maximum Oxygen Requirements

Oxygen	$= 0.76\ (BOD_r) + 4.33\ (N_r) + 0.21\ (MLVSS)(x_d)$
BOD_r	= Need to know BOD removal by anoxic selector (assume 90%)

N_r = TKN Oxidized = 4,000 – 0.04 (15,000)
= 3,400 lbs/day
MLVSS = 1.29 MG × 4,500 mg/L × 8.34 = 48,415 pounds
B^1 @ 30°C = 0.21/day
X_d (degradable MLVSS) = 0.5 (Assume)

Assume anoxic selector in designed for 90% BOD_r

BOD to Aeration Basin = 15,000 × 0.1 = 1,500 lbs/day
Oxygen = 0.76 (1,500) + 4.33 (3,400) + 0.21 (48,415) 0.5
= 1,140 + 14,722 + 4,842
= 20,704 lbs/day

Assume maximum oxygen transfer rate or oxygen uptake rate of 125 mg/l/hr *(Always confirm with aeration equipment vendor)*

Aeration Basin Volume = 20,704/125 × 24 × 8.34 = 0.83 MG
= 0.83 MG vs. 1.29 for Nitrifiers

Therefore, nitrification controls over aeration in Option 1.

Anoxic Volume

$t = (S_o - S_e)/f_a X_v K\ (S_e/S_o)$
$K_{30°C}$ = 15/day
S_o = 3,600 mg/L BOD
S_e = anoxic basin effluent, mg/L
X_v = MLVSS, 4,500 mg/L
f_a = 0.70 active fraction of biomass
t = detention time in selector, days

Calculate detention time (t) for various % BOD removal and effluent BOD concentrations (S_e) from the selector.

Select anoxic reactor volume of 345,000 gallons for further refinement and optimization using the Biowin Model and determining the MLSS recycle rate back to the selector to provide nitrate-N for BOD removal.

% BOD Removal	S_e (mg/L)	t (days)	Anoxic Basin Volume (gallons)
85	540	0.43	215,000
90	360	0.69	345,000
95	180	1.45	725,000

Option 2—Convert MLE Activated Sludge to MBR Processes

The main difference in the MBR alternative is that higher MLSS and nitrifier biomass concentrations can be carried in the anoxic reactors and aeration basins. What design MLSS should be selected? It depends on the type of membrane system and needs to be confirmed with the vendors.

Assume 12,500 mg/L MLSS for a flat plate membrane system for this example.

Nitrification needed:

Assume 12 days SRT and 7.0% nitrifier fraction of the MLVSS

Nitrifiers = 12,500 (0.9)(0.07)

Nitrifiers = 788 mg/L (versus 315 mg/L for MLE)

Aeration Basin Volume = 3,400/788 × 8.34

= 0.52 MG

Check aeration basin volume needed to transfer the oxygen required for BOD removal and nitrification.

Assume Oxygen Required = 20,704 lbs/day (same as MLE)

Assume maximum oxygen transfer rate of 125 mg/l/hr (Always confirm with aeration equipment vendor)

Aeration Basin Volume = 0.83 MG > 0.52 MG (1.66 times existing volume)

Therefore, oxygen controls total aeration basin volume in Option 2.

Adding membrane cassettes to existing clarifier may be able to satisfy the total aeration basin volume required.

In addition to comparing anoxic and aeration treatment tank volumes, a comparison of secondary clarifier requirements and membrane tank requirements must be included in the process comparison.

Option 3—Convert MLE Activated Sludge to BioMag

This is another option for increasing the MLVSS biomass concentration in the aeration basins to potentially eliminate the need for additional aeration basin volume. Several facilities are in operation carrying up to 10,000 mg/L MLVSS using magnetite with specific gravity of 5.2 to increase the solids settling rate in the final clarifier.

REFERENCES

1. Hydroscience Report, 1978.
2. Blum, J.W. and R.E. Speece. 1990. "A Database of Chemical Toxicity to Environmental Bacteria and Its Use in Interspecies Comparisons and Correlations." Vanderbilt University.
3. Sutton, P.M., T.R. Bridle, W.K. Bedford and J. Arnold. 1979. "Nitrification and Denitrification of an Industrial Wastewater." First Workshop, Canadian-German Cooperation Water Pollution Control for the 80's. Wastewater Technology Center, Burlington, Ontario, Canada.
4. Eckenfelder, Jr., W. Wesley, Davis L. Ford and Andrew J. Englande, Jr. 2008. *Industrial Water Quality*. Fourth Edition. McGraw Hill.

CHAPTER 6

Microconstituents Removal

ACTIVATED SLUDGE has been proven to be very effective for removal of microconstituents and specific organics in both municipal and industrial wastewater treatment plants. Over the last several years, the authors have been involved with treatability testing, process modeling and evaluation of the fate of microconstituents such as pharmaceutical actives and insecticides through activated sludge plants. Activated sludge has been shown to be a key treatment process to remove the bulk of these microconstituents through biodegradation and adsorption to the biomass. Additional polishing treatment can be added to achieve lower effluent concentrations and reduce effluent toxicity and estrogenicity effects. The polishing treatment technologies that are typically used include: chemical oxidation, membranes, carbon and ultrasound.

Microconstituents including active pharmaceutical ingredients (APIs) are becoming a concern and are being evaluated by EPA for future regulations. EPA has already issued Final Water Quality Criteria for nonylphenol [1]. Some industries, such as pharmaceutical, have been proactive and are already implementing treatment to control microconstituents entering the environment. Municipalities are studying the removal of microconstituents through treatment processes including what is contained in biosolids for land application and composting.

Microconstituents, the term developed by the Water Environment Federation [2], can make their way into the environment through a variety of routes, such as industrial discharges, wastewater treatment plant (WWTP) effluent, runoff from agricultural and feedlot operations, and other nonpoint sources that are more difficult to quantify. Microconstituents are defined as natural and manmade substances, including elements and inorganic and organic chemicals, detected within water and the environment for which a prudent course of action is suggested for the continued assessment of the potential impact on human health and the environment. Humans, aquatic organisms, and other wildlife can be exposed to these compounds through environmental exposure and consumption of certain foods and waters.

Wastewater treatment plants serve as a collection point for compounds originating from or used in residential, commercial, and industrial applications; surface runoff is also included in combined sewer systems. Depending on the processes in place at the WWTP and the physical and chemical characteristics of the wastewater, some of the influent compounds that are not currently regulated may not be removed or may be only partially degraded through treatment. For this reason, many studies around the world [2] have focused on the endocrine disruption observed in various species of aquatic organisms in the vicinity of WWTP outfalls. These WWTPs have been the focus of research because they represent a seemingly straightforward point-source to target for investigation, and not necessarily because they are the most important, or significant sources of microconstituents in the environment. Compounds that have most often been implicated in endocrine disruption in aquatic organisms are the natural estrogens estrone (E1) and estradiol (E2), which are excreted by all humans; the synthetic estrogen ethinylestradiol (EE2), which is the active ingredient in birth control pills; and nonylphenol and octylphenol. It should be noted that "safe" environmental levels of these compounds have not been agreed on by the scientific community; additionally, conclusions may change as research continues and the data base broadens [2].

Microconstituents include a few sub-groups, one of which is Pharmaceuticals and Personal Care Products (PPCPs). The PPCPs includes a large number of chemical contaminants. These can originate from human usage and excretion and veterinary applications. A variety of these products, such as prescription/non-medications, fungicides and disinfectants are used for industrial, domestic, agricultural and livestock practices (Ternes and Daughton, 1999)[3].

Figure 6.1 illustrates that endocrine disruptors include some but not all human/veterinary drugs such as synthetic hormones, pesticides such as DDT and lindane and industrial chemical such as bisphenol A and nonylphenol (Kobylinski *et al.*, 2005)[4]. Not all PPCPs, PhACs, or microconstituents are EDCs. Due to significant advances in analytical technologies, microconstituents are being detected in the environment; however, the fact that a compound is detected as a microconstituent does not necessarily mean that it is harmful or detrimental to the environment.

Table 6.1 shows the diversity of endocrine disruptor compounds including pesticides and herbicides, persistent industrial chemicals, heavy metals and hormones (Jovic, 2004)[5].

There are also potential EDC's that might be expected to lead to endocrine disruption in an intact organism. These potential EDCs include: hormones, natural chemicals, synthetic pharmaceuticals and other man-made chemicals including cosmetics and pesticides. All of the potential EDC's that are cur-

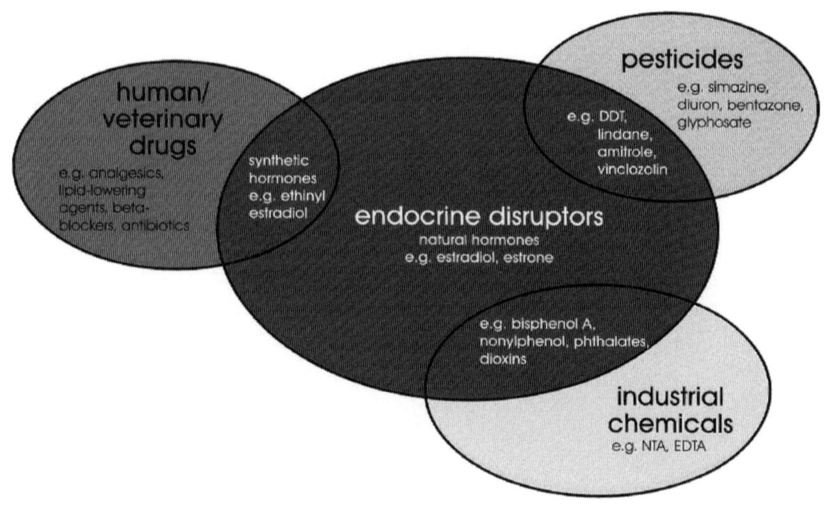

Figure 6.1. Endocrine disruptor compounds [4].

rently recognized probably form only a small proportion of the total in the environment (Jovic, 2004)[5].

EDC's have received more attention in the media and technical literature in recent years. A review by the World Health Organization [6] has concluded that low-level environmental exposure to Endocrine Disruptors has not yet been demonstrated to cause harm to human health (Damstra *et al.*, 2002)[7]. Uptake of Endocrine Disruptors by humans from treated drinking water is relatively low in comparison to other sources such as food (Global Water Research Coalition, 2003)[8].

TABLE 6.1. Diversity of Endocrine Disruptor Compounds [5].

Pesticides & Herbicides	2,4-D Endrin Vinclozolin Lindane	Chlordane Heptachlor Atrazine	DDT Methoxychlor Simazine	Endosulfan Toxaphane Dieldrinlagooning
Persistent Industrial Chemicals	Dioxins PCP Phthalates	Furans Hexachlorobenzene Bisphenol A	P-nonylphenol's Tributyl tin	PBB Octachlorostyrene PCB's
Heavy Metals	Cadmium	Lead	Mercury	
Hormones	Synthetic: Biological: Plant secondary metabolites	17α-Ethinylestradiol 17β-Estradiol Progesterone Lignans	Diethylstilbestrol Estriol Testerone Sesquiterpenes Isoflavonoids	Estrone Phytosterols Coumestans

PHYSICAL/CHEMICAL PROPERTIES

The fate of microconstituents including the PPCPs and EDCs through municipal and industrial activated sludge treatment plants is driven mainly by their physical/chemical properties. Pharmaceuticals are complex molecules with different physical and chemical characteristics. They are notably characterized by their ionic nature, molecular weights between 300 and 1000, and relatively low solubility (mg/L or less). Most pharmaceuticals exist as cations, anions, or zwitterions, with only a limited number of compounds existing in neutral forms under the conditions encountered in the aquatic environment. They are designed to have specific effects within humans (for example, cardiovascular, central nervous system, antibacterial, anti-inflammatory, immune, and anticancer). Many of these special properties are relevant to the assessment of their environmental fate and transport properties (Williams, 2006)[9].

Physical and chemical properties describe the forms of the compounds and their partitioning into various environmental compartments. The major physical and chemical properties affecting environmental fate and transport of pharmaceuticals are: (Williams, 2006)[9].

- Water solubility, S
- Dissociation constant, pK_a
- Octanol and water partition coefficient, K_{ow}
- Vapor pressure, Vp
- Distribution coefficient (K_d), which sometimes is expressed as the organic carbon-normalized distribution coefficient (K_{oc}).

The key processes that characterize the rate of transformation of organic contaminants in the environment are:

- Biotransformation rate (k_{bio}) in soil, water, sludge, or sediment
- Photolysis rate, k_{photo}
- Hydrolysis rate, k_{hydro}
- Oxidation rate (via a specific oxidant)
- Reduction rate (via a specific reductant)

APPLICABLE TREATMENT PROCESSES

Deactivation of APIs using sodium hydroxide or caustic at high pH of about 10 and sodium hypochorite is a common practice used to deactivate or destroy APIs. This technology has been used for antifungal compounds, antibiotics, cancer drugs, mood stabilizers and others. Detention times are typically up to 24 hours for treatment of the high strength product solution remaining in the tank and/or piping prior to dilution with any other wastewaters. Typical concentrations of APIs are 10,000 to 50,000 mg/L prior to any dilution. One concern is whether the deactivation results in breakdown products which could also be harmful in downstream treatment processes or to aquatic life if they pass through the treatment plant.

Biodegradation in the activated sludge process often is very effective at removal of microconstituents to low concentrations.

Sorption is an important process that controls the distribution of pharmaceuticals between particles and water. Charged compounds, such as many pharmaceuticals, typically undergo sorption by mechanisms that are different from those of neutral hydrophobic compounds. As a result, prediction of pharmaceuticals partitioning based on only hydrophobic interactions, K_{ow}, will tend to overestimate sorption. On the other hand, distribution coefficients, D_{ow}, which accounts for partitioning of both unionized (hydrophobic) and ionized (hydrophilic) forms of a molecule, tend to underestimate sorption.

The adsorption of pharmaceuticals to suspended particles or biomass in the activated sludge process represents a potentially important attenuation mechanism for many human pharmaceuticals in municipal WWTPs. Adsorption of organic compounds onto surfaces usually is considered to be a reversible equilibrium process that depends upon the nature of the compound, the surface, and the solution composition (Williams, 2006)[9]. There are currently no correlations of the observed sorption K_d values with literature values (i.e., octonal water partitioning K_{ow} or partitioning to soil K_{oc}) except for musk fragrances. Therefore, the sorption coefficient has to be measured for each compound and for each sludge type (Ternes *et al.*, 2005)[10].

The European Union (EU) Poseidon Project established a basic knowledge of the removal of PPCPs and estrogens in wastewater and drinking water systems. In the Poseidon project, Ternes *et al.* (2005)[12] studied ibuprofen, estrogen, antibiotics, heavily sorbing musk fragrances persistent compounds such as carbamazepine and biodegradable natural estrogens. Treatment technologies tested included membrane technology and effluent ozonation in combination with activated sludge and biofilter systems (for nitrification an denitrification). Sludge treatment technologies as well as novel sustainable approaches such as urine separation were evaluated. The main conclusion was that biological degradation and sorption are the key mechanisms for PPCP removal during municipal wastewater treatment. Sorption onto suspended solids is an important mechanism for hydrophobic and positively charged compounds. Biodegradation can be described as pseudo first order kinetics. Stripping was not effective except for some of the volatile musk fragrances. Chemical oxidation was an effective polishing step for some PPCPs. Some of the PPCPs are degraded significantly in anaerobic sludge digestion.

For most PPCPs, the MBR is comparable to activated sludge run at comparable sludge age (Ternes *et al.*, 2005)[10]. A biofilter (fixed bed reactor) with nitrogen removal performed comparably to an activated sludge reactor in parallel indicating shorter hydraulic residence times are compensated by higher bioactive sludge concentration in fixed film system (Ternes *et al.*, 2005)[10].

EDC removal is being studied at an MLE plant in San Diego, California and at a MBR pilot plant in New Mexico as part of a Water Environment Research Foundation (WERF) study (Oppenheimer, 2005)[11]. Oppenheimer *et al.* (2006)[12] reported that the research study for WERF evaluated removal of PPCP in six full-scale wastewater treatment plants in the United States and two pilot-scale MBRs operating over a wide range of SRT values. Study results showed that half of the 20 PPCP were removed at greater than 80% at a critical SRT of less than five days. A small group of PPCPs required higher SRT to 15 days for removal at greater than 80%. From the limited data, no additional removal could be attributed to the use of MBRs, media filters or longer hydraulic retention times. Removal of musk fragrances required SRTs of at least 25 days.

Esposito *et al.* reported (Esposito, 2007)[13] on removals of microconstituents through pilot studies in Logan Township, New Jersey. The treatment system included activated sludge followed by reverse osmosis, and UV/hydrogen peroxide. The treatment effluent showed non-detectable estrogenic activity. The treated effluent was comparable to that of the groundwater it would be injected into for indirect potable reuse. The treatment system was referred to as "multiplier barrier approach" and the treated municipal wastewater was to be used for indirect potable reuse.

Miroconstituent removal through the Kalamazoo, Michigan POTW was reported by EPA (EPA, 2007)[14]. This 24 mgd plant has the PACT process which is essentially activated sludge with the addition of powdered activated carbon to the aeration basin. The removal of numerous microconstituents ranged from 75 to 99.9%.

Some EDCs are theoretically fully biodegradable. Johnson and Darton (2003)[15] state that the EDCs generally implicated in endocrine disruption in fish (i.e., the estrogens E1, E2, EE2, and nonylphenol) are all "inherently biodegradable and so in theory should not present an intractable problem." Many studies have demonstrated that some activated sludge processes have the potential to remove a large fraction of several suspected EDCs, often to below detection limits (Ying *et al.*, 2002)[16]. The amount of EDC removal achieved depends on both the characteristics of the wastewater and the processes in place at the WWTP.

Researchers have found that solids retention time (SRT) in an activated sludge system has a pronounced effect on removal of some EDCs (Joss *et al.*, 2004)[17]. Solids retention time has a major influence on biological degradation efficiency, which is mainly due to its influence on the diversity of the microbial population and on the multitude of degradation pathways being expressed. Philips *et al.* (2005)[18] noticed marked improvement in EDC removal as SRT increased from 5 to 10 days. It appears that SRTs of 10 to 15 days may result in more effective removals (Seigrist *et al.*, 2005)[19]. Each WWTP and wastewater is unique, however, so the degree of EDC removal will be site-specific.

Case Study No. 1—Removal of Pharmaceutical Actives in Full-Scale Pharmaceutical Treatment Plants

This case study involved the evaluation of full-scale plant sampling of four microconstituent compounds at two activated sludge plants in Europe. The physical/chemical characterization was input to a process model to compare predicted performance versus actual full-scale performance. Two of the compounds were highly biodegradable and one, an antifungal compound, was not. Partitioning or adsorption to sludge based on the octanol-water coefficients was minimal which correlated well with the full-scale performance. The model was also used to predict the increase in removal of one compound with increased sludge age from 15 to 30 days. The removal of the same API at another plant was shown to be 95% at higher sludge age is shown in Table 6.2. A feasibility study was performed to evaluate various alternatives to upgrade the two plants to increase API removal. This evaluation upgrading activated sludge treatment plus adding reverse osmosis and advanced oxidation processes. PACT was also evaluated. All of these alternatives were evaluated in combination with source control.

The STPFATE9 model (Mueller, 1995)[20] was used to determine the impact of increasing the sludge age on the fate of through one of the treatment plants. Based on measured data of the concentrations of these organics in the influent, effluent and sludge, a previous evaluation determined that for the major removal mechanism was biodegradation with negligible removals occurring through sorption. Volatilization losses were assumed to be zero. The biodegradation losses were computed to bc approximately 66 and 73 percent for the two compounds. For the STPFATE9 analysis, the treatment plant was

TABLE 6.2. Example Mass Balance.

	Plant 1	Plant 2
Influent API (μg/l)	32.6	13
Effluent API (μg/l)	11	0.1
Sludge API (μg/l)	2	0.8
% Effluent	34	< 1
% Adsorbed to Sludge	< 1	< 1
% Volatilized (assumption)	0	0
% Biodegraded	66	99
Sludge Age (days)	15	29
Aeration Basin Temperature (°C)	26	30

simulated by setting the plant's physical dimensions and operating conditions (i.e., BOD removals, flow, solids, tank volumes, etc). The measured influent concentrations, chemical-specific octanol-water partition coefficient and biodegradation rate constants were inputted to simulate the measured sludge concentrations (i.e., sorption losses) and biodegradation losses (i.e., computed values). The plant was calibrated to a 15 day SRT, which is the condition the plant is currently operating under. Once all the conditions were set, the SRT was increased to 30 days and the impact on the biodegradation was determined. The result was an increase in the biodegradation rate by 14%. The percent removed by biodegradation increased from 66% to 75% for one compound and 73% to 83.5% for the second compound.

Case Study No. 2—Full Scale Plant with MBR + Ozone

Advanced oxidation processes are combinations of ultraviolet light (UV) plus hydrogen peroxide, ozone plus hydrogen peroxide, and UV plus ozone that are specifically designed to increase the formation of powerful hydroxyl radicals to oxidize microconstituents. Ozone shows similar potential as Advanced Oxidation Processes (AQPs) for microconstituent removal and is often included in AOP discussions. Substances that are difficult to biodegrade may be oxidized by AOPs, and the oxidized byproducts may be more amenable to biodegradation. Advanced oxidation processes can be followed by a biological process to further degrade the byproducts; or, natural purification processes may be relied upon for treatment, depending on the situation (Ried and Mielcke, 2003)[21].

Table 6.3 shows the results of a two-stage MBR and ozone treatment pilot study (Helmig, 2009)[22]. These results show removal of active pharmaceutical ingredients (APIs) through MBRs but still some residual concentration which was further destroyed by ozone treatment to non-defect concentrations. Estrone was still present in the effluent after ozone treatment.

Full scale performance for this MBR plus ozone treatment plant showed a 1,000 fold reduction in estrogenicity using Microtox™ test as 17β-estradiol equivalents (Helmig, 2009)[22]. The total APIs removal was 99.94%. The full scale operating plant demonstrates that these technologies are proven at full scale. Figure 6.2 shows the process flow diagram of the full scale facility. Figures 6.3 and 6.4 show the API and estrogenicity removals, respectively. Figure 6.5 shows a picture of the facility.

A WEFTEC Workshop in 2011[23] by Cleary, Helmig and Grey covered the technologies that are working for removal of pharmaceutical activities including activated sludge. Chapter 7 presents an approach for alternatives analysis for removing microconstituents. The recommended approach along with case studies includes the use of treatability studies and process modeling tools.

TABLE 6.3. Pilot Scale Results—MBR Performance—APIs.

API	Units	Detection Limit	Influent	Permeate	Reduction (%)
17-α-estradiol	μg/L	0.23	0.36	0.18	> 50.0
17-β-estradiol	μg/L	0.23	1.73	0.30	82.4
17-α-dihydroequilin	μg/L	0.23	1.66	0.30	81.7
Estrone	μg/L	0.23	3.85	0.30	14.3
Trimegestone	μg/L	0.87	8.13	0.43	> 94.7
Ethinyl estradiol	μg/L	0.03	54.8	17.6	67.9
Estriol	μg/L	7.93	52.9	4.0	> 92.5
Medrogestone	μg/L	7.90	11.6	4.0	> 65.9
Norgestrel	μg/L	0.87	52.0	2.7	94.9
Estradiol valerate	μg/L	0.87	20.4	2.8	86.1
SSRI (Venlafaxine)	mg/L	0.05	2.80	2.12	24.3

A recent project called "Pharmaceutical Input and Elimination from Local Sources (Pills)" is being conducted by a European partnership led by Emschergenossenschaft, a German Water Board[24]. This work, which started in 2007, continues to evaluate new treatment technologies in hospital wastewater to evaluate removal of APIs and increase awareness of APIs across Europe. The study is evaluating: (1) which activities in pharmaceutical wastewaters have ecotoxicological effects and therefore may therefore present the greatest risk to the environment; and (2) which activities are not removed by advanced treatment technologies. Pilot plant studies are being conducted at hospitals in four countries (i.e., Switzerland, Germany, The Netherlands and Luxembourg). The technologies being studied include primary treatment followed by

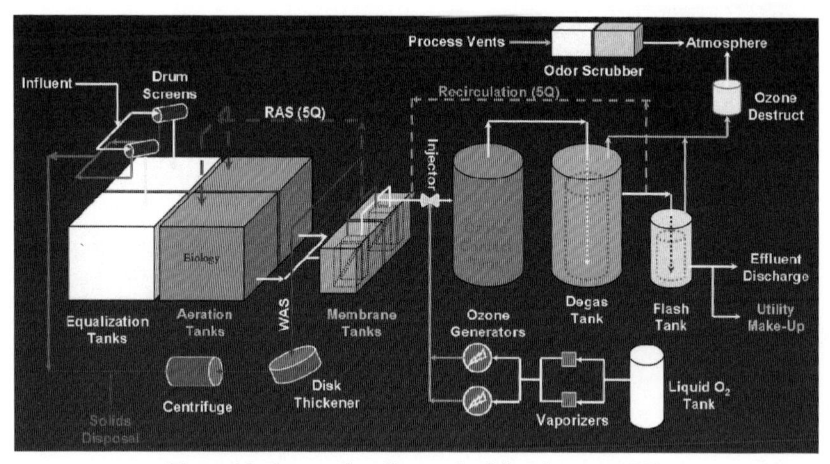

Figure 6.2. Process flow diagram for full scale system [22].

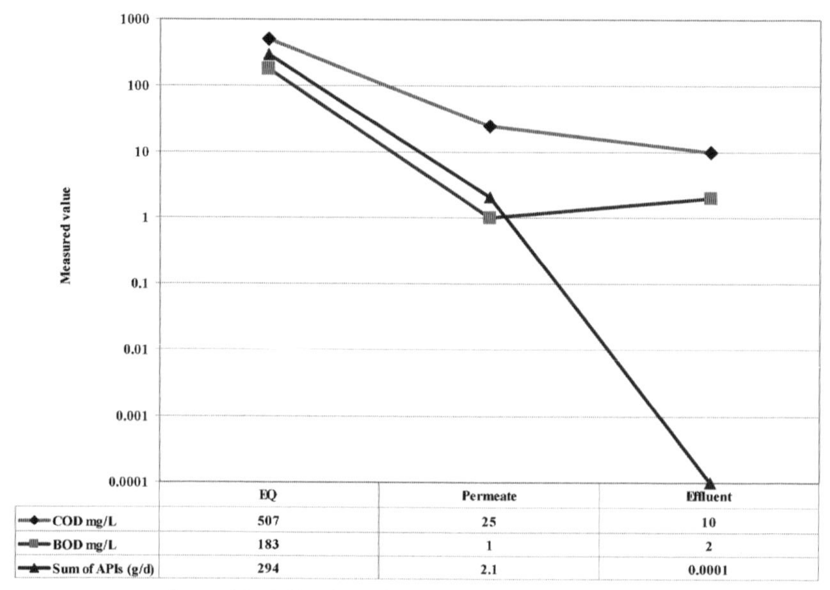

	EQ	Permeate	Effluent
COD mg/L	507	25	10
BOD mg/L	183	1	2
Sum of APIs (g/d)	294	2.1	0.0001

Figure 6.3. Full scale results—APIs and gross organics [22].

MBR followed by either PAC, ozone or Titanium Dioxide/UV, followed by sand filtration.

Raczko, R. of United Water, recently presented at NJWEA Technology Transfer Workshop (March 2013)[25] on pilot studies conducted at their water treatment plant in Haworth, New Jersey, as well as sampling of EDCs

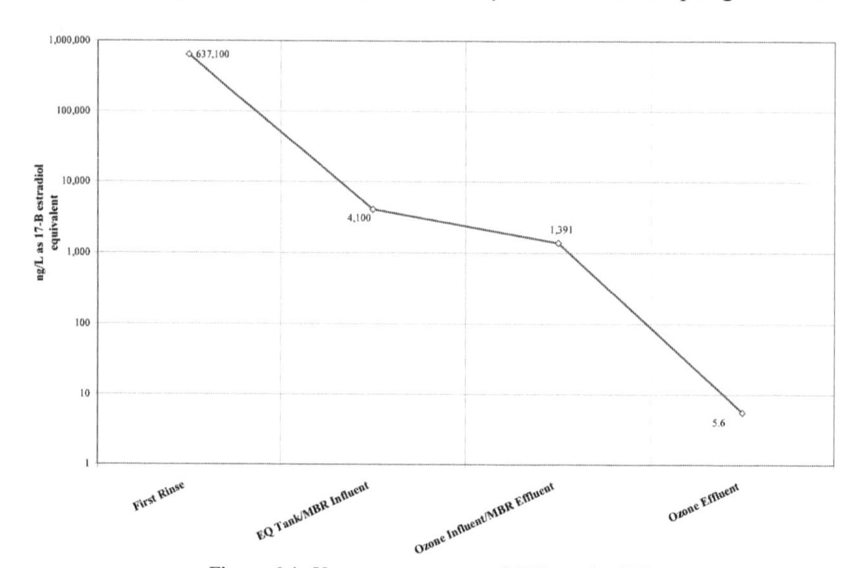

Figure 6.4. Yeast estrogen screen (YES) results [22].

Figure 6.5. Full scale treatment system [23].

through full-size water treatment plants. This research project was a collaboration of Water Research Foundation (Water RF), United Water, Black & Veatch, NJDEP, PVWC and NJIT. The purpose of the study was to investigate the effectiveness of full-scale conventional and advanced water treated processes for removal of EDCs. Twelve classes of compounds, which included sixteen individual compounds, were the priority indicator compounds. Pharmaceutical and personal care products were defeated in the raw water including: pharmaceuticals, antibiotics, flame retardants, flavor and fragrance, PAHs, pesticides, solvent, plasticizers, and detergents (nonyl and octylphenols).

Advanced oxidation was effective for oxidation/conversion of many aromatics, alkanes, and cyclic organics. GAC adsorption was effective for the more non-polar compounds (PAHs, flavor and fragrance compounds, and many of the pharmaceuticals). Intermediate ozone and virgin GAC was the most effective treatment combination to achieve non-deterrent effluent concentration for 13 of the 16 compounds. Future work was recommended to determine the effective dosage for ozone and hydrogen peroxide, the detention time and usage rate for GAC and for potential by-products formed.

As assessment was made of the ability of wastewater treatment plants to remove nonylphenol ethoxylate (NPE) surfactants that are used widely in domestic and industrial applications[26]. This work was in preparation for the publication of U.S. Environmental Protection Agency (EPA). Water Quality Criteria for nonylphenol, a biodegradation intermediate of nonylphenol ethoxylates. As with the degradation of many anthropogenic compounds, it is important to operate biological treatment systems at high solids retention times

(SRT) with minimum levels of approximately 10 days. This observation has been confirmed by the results from more recent investigations.

Nonylphenol (NP) is unique among the commonly discussed microconstituents because it is the subject of regulation: under Section 304(a)(1) of the Clean Water Act and also, because the EPA has finalized federal ambient aquatic life Water Quality Criteria (WQC)[1]. NP is a biodegradation intermediate of nonylphenol ethoxylate surfactants, which are of interest because of their use in consumer and industrial application and the common occurrence of their degradation intermediates in wastewater.

The physical-chemical properties of NPEs and their degradation intermediates can provide insight into the potential behavior of these substances in wastewater treatment plants and the environment. For example, octanol-water partition coefficients (log K_{ow}) are useful for assessing the partitioning of compounds between environmental media. Partition coefficients for higher mole NPEs are relatively low (~3.0) but higher for the low mole NPEs and NP (3.3 to 4.4). The higher mole NPEs will tend to stay in the water phase as opposed to becoming associated with sediments, which is consistent with the reported literature.

High mole NPEs are quite soluble in water (> 10,000 mg/L), while the lower mole NPEs and NP are much less soluble (5 to 50 mg/L). It may, therefore, be expected that the low mole NPEs, if released to water, would be subject to volatilization into air. Based on their low vapor pressures (< 0.001 Pa) and high water solubility, partitioning of high mole NPEs to air would be negligible (Henry's law constants, H_c < 1E-4 Pa m^3/mol). In contrast, NP is more volatile, with H_c values in the range of 1 to 10 Pa m^3/mol.

Many studies showed that the solids retention time (SRT) was the most influential of the treatment plant operating parameters for the control of NPEs. A value of approximately 10 days or greater appeared to be appropriate to ensure high removals of NPEs.

One of the most comprehensive studies supporting this conclusion is that of Johnson [27] who studied the fate of NP, a NPE metabolite. NP effluent concentrations were measured at 14 municipal wastewater treatment plants in eight European countries. Figure 6.6 summarizes their data. NP is a suitable surrogate for examining the influence of operating conditions on the removal of NPEs given that rapid primary degradation of the ethoxylate chain is known to occur. All of the plants examined utilized activated sludge systems modified to incorporate biological nutrient removal (BNR). Unfortunately, no influent data were reported so the true effect of SRT could not be evaluated with respect to specific NP loading, nor was it possible to estimate NP removal efficiencies. They could not find any statistically significant correlation between NP effluent concentration and HRT and SRT at the 10 percent level. This is probably because the comparison is being made between plants that are all employing state-of-the-art treatment technology and the effluent NP concentrations were all very low (< 2 μg/L).

Martz, Manfred (2012)[28] presented a strategy to effectively treat pharmaceutical wastewater. He states that API molecules are mostly very stable and have a high persistence for biodegradation and most effectively removal in converted wastewater treatment plants. The article states that the preferred solution for APIs is to treat at the source by deactivation prior to dilution with other wastewaters. Physical and chemical properties and potential treatment processes were presented for 17α-Ethinylestradial (EE2) an oral contraceptive active and x-ray iodized contrast media. The EE2 is biodegradable under optimized conditions and adsorbable to biomass with a K_{ow} of 4.2 at pH 7. The x-ray CM is hard to biodegrade and has low adsorption. Ozonation and carbon adsorption treatment processes have been effective on the x-ray CM at up to 75 to 82% removal.

A study by WERF [29] investigated Trace Organic Compounds including pharmaceuticals and personal care products during conventional wastewater treatment. Twenty-two compounds were selected for the study on the fate during conventional treatment and to assess their suitability to serve as potential performance indicators. The removal of the compounds was studied on bench-scale, pilot-scale and full-scale treatment plants. Biotransformation and sorption characteristics were quantified in controlled laboratory experiments to support process modeling using AS Treat model. The 22 compounds were grouped into nine categories in a 3 by 3 matrix using low, medium and highly biodegradable and low, medium and high sorption as shown in Table 6.4.

The indicators selected exhibited a high detection ratio (> 10) and detection frequency in wastewater influents. Only two of the targeted indictor com-

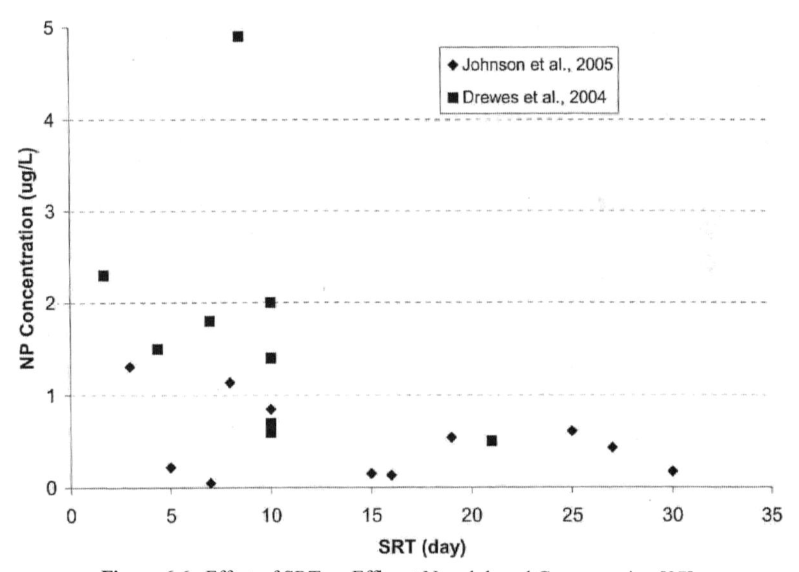

Figure 6.6. Effect of SRT on Effluent Nonylphenol Concentration [27].

pounds, DEET and caffeine, quantified in wastewater influent concentrations at different facilities throughout the U.S. exhibited a strong seasonal and regional dependency.

The efficiency and mechanisms of TOrC removal were evaluated during full-scale activated sludge treatment under steady-state process conditions. Full-scale sampling was conducted at seven wastewater facilities in the U.S. during 13 independent sampling campaigns. This resulted in detailed TOrC mass balances primarily around the secondary treatment processes quantifying removal by sorption and biotransformation for each TOrC indicator. The selected facilities used Conventional Activated Sludge (CAS), High Purity Oxygen (HPO), Modified Ludzack-Ettinger (MLE), Biological Nutrient Removal (BNR), and Membrane Bioreactor (MBR) processes. Plant influent flows ranged from less than 1 mgd to over 90 mgd and operating SRTs from less than two to over 50 days.

The observed TOrC removal during secondary treatment could be linked to the bin categories established for the indicators on the basis of sorption and biotransformation properties measured in the mixed liquor of various wastewater treatment facilities as shown in Table 6.5. The measured sorption and biotransformation characteristics were predictive of the removal efficiencies for the majority of TOrC indicators during full-scale secondary treatment. Table 6.5 summarizes the anticipated removal efficiencies of TOrC indicators during activated sludge treatment based on the nine bins. It is anticipated that similar efficiencies will be achieved for other TOrC that fall into the respective bin grouping based on their biotransformation and sorption characteristics.

Several indicators were not effectively removed during secondary treatment independent of process operation (less than 30% reduction). They included TOrC that were low in sorption and biotransformation potential (i.e., carba-

TABLE 6.4. Summary Matrix of TOrC Indicators Based on Biotransformation and Sorption Fate Parameters [29].

		Biotransformation (k_b, L/g-d)		
		Slow < 0.1	**Moderate 0.1–10**	**Rapid > 10**
Sorption (log K_d)	Low < 2.5	Carbamazepine Meprobamate Primidone TCEP Sucralose	DEET Sulfamethoxazole Gemfibrozil Iopromide Trimethoprim	Acetaminophen Caffeine Naproxen Ibuprofen Atenolol
	Moderate 2.5–3	TCPP	Cimetidine	Benzophenone Diphenhydramine Bisphenol A
	High > 3	Triclocarban		Triclosan Fluoxetine

TABLE 6.5. Anticipated Overall Removal of TOrC Based on Biotransformation and Sorption Characteristics [29].

		Biotransformation (k_b, L/g-d)		
		Slow < 0.1	**Moderate 0.1–10**	**Rapid > 10**
Sorption (log K_d)	Low < 2.5	0–30% (Typical: 5%)	0–100% (Typical: 70–90%)	70–100% (Typical: 95%)
	Moderate 2.5–3	0–60% (Typical: 20%)	0–100%(2) (Typical 30-50%)	60–100% (Typical: 70%)
	High > 3	0–95% (Typical 50%)	N.A.	0–100%(1)

(1)Data basis weak to estimate removal for this group.
(2)The anticipated removal can be narrowed for a specific compound and process operation by using the threshold SRT 80% identified in this study.
N.A. = not available.

mazepine, primidone, TCEP, and sucralose). Advanced wastewater treatment is required to effectively attenuate these types of compounds. Implementation of advanced treatment processes is anticipated to also enhance the removal of TOrC in other bin categories.

Solid retention time, hydraulic retention time, wastewater temperature, solids recycles, redox conditions, overall process stability, and TOrC influent concentrations were important factors affecting the removal efficiency of TOrC through sorption and biotransformation. These relationships could be quantified in this study and are helpful to predict TOrC removal during conventional treatment on basis of process configuration and operational boundary conditions.

Facilities that operated at long SRTs demonstrated generally higher removal efficiencies for TOrC that are amenable to biotransformation than facilities operating at short SRTs. This finding indicates treatment synergies between TOrC removal and nitrification for facilities that are operating at a high level of secondary treatment to meet low ammonia limits. Conventional secondary treatment does not provide a significant barrier against TOrC that fall into the bin slow biotransformation and low sorption. Removal of these compounds requires advanced treatment processes beyond conventional wastewater treatment.

Findings of this study [29] did not indicate that the addition of anoxic conditions during secondary treatment at facilities operating in denitrification mode improved the removal of TOrC indicators. Several moderately biotransformed compounds showed an increase in liquid phase concentration after anoxic treatment before being removed in the aerobic portion of the aeration basins. Thus, anoxic conditions did not compromise the overall treatment efficiency for TOrC during secondary treatment. The increase in liquid phase concentration after anoxic treatment could have been caused by desorption of solid bound TOrC under oxygen deficient conditions. Similar desorption processes may also occur when solids are temporarily stored in secondary clarifiers and may lead to temporary increased TOrC concentrations in secondary effluents.

TOrC modeling was conducted in this study [29] using ASTreat, due to its free access, simplicity, and suitability from a utility perspective regarding easily available input parameters. The fundamental effort of this study to develop indicator-specific fate parameters for sorption and biotransformation provides the necessary basis for the application and evaluation of other TOrC fate models that could not be considered within the scope of this study. The accuracy and reliability of TOrC fate modeling was improved by determining accurate compound-specific biotransformation rate parameters and sorption coefficients as model inputs. The library of fate parameters developed in this study can give guidance for selecting appropriate biotransformation rate constants and sorption coefficients for the TOrC indicators for future use based on general activated sludge process conditions.

ASTreat proved to be a useful screening tool for predicting the removal of most TOrC indicators under full-scale treatment. The accuracy of predicting the removal for TOrC that are moderately fast biotransformed was improved by recognizing that TOrC biotransformation rates are a function of the operating SRT. The fate prediction of TOrC that are sorbable and rapidly biotransformed remains a major challenge, as these compounds appear to accumulate

TABLE 6.6. Threshold SRT Values to Achieve at Least 80% TOrC Removal [29].

	SRT, days
Acetaminophen	2
Caffeine	2
Ibuprofen	5
Naproxen	5
Bisphenol A	10
Triclosan	10
DEET	15
Gemfibrozil	15
Atenolol	15
BHA	15
Iopromide	15
Cimetidine	15
Diphenhydramine	20
Benzophenone	20
Trimethoprim	30

Threshold SRT values could not be determined for fluoxetine, TCPP, TCEP, primidone, sulfamethoxazole, carbamazepine, triclocarban, or sucralose because compounds are recalcitrant or removal was too variable.

on the solids during treatment, making a steady-state performance analysis, as attempted in this study, challenging.

Laboratory-scale flow-through experiments were performed to systematically assess the effect of SRT on TOrC removal. Three systems were operated in parallel at 5, 10, and 20 days SRT treating the same feed water. The HRT for all three systems was kept constant at ~20 hours. Four weekly sets of samples were analyzed for TOrC removal. TOrC removal was also assessed through a pilot-scale sequencing membrane bioreactor (SMBR), which treated the same wastewater source as the flow through systems. The results from both experiments generally confirm the effect on SRT on TOrC indicator removal as shown in Table 6.6.

REFERENCES

1. Aquatic Life Ambient. Water Quality Criteria—Nonylphenol. Final U.S. E.P.A., EPA-822-R-05-005. December 2005.
2. Water Environment Federation. 2007. Technical Practice Update "Effects of Wastewater Treatment on Compounds of Potential Concern".
3. Ternes, T.A. and C.G. Daughton. 1999. "Behavior and Occurrence of Estrogens in Municipal Sewage Treatment Plants." *The Science of the Total Environment, 225*, 81–90 and 91–99.
4. Kobylinski, H. and Scruggs, C. 2005. "Are We Ready to Begin Permitting EDCs?" WEFTEC 2005. Proceedings, Washington, D.C., October 29 to November 2, 2005.
5. Jovic, A. 2004. "Technical Aspects of Endocrine Disruptor Degradation," Endocrine Disruption Workshop, Melbourne, July, 2004, Victorian Water Industry Association and Deakin University.
6. The World Health Organization. "The International Program on Chemical Safety".
7. Damstra *et al.* 2002. "Global Assessment of the State-of-the-Science of Endocrine Disruptors."
8. Global Water Research Coalition, UK. 2003. "Endocrine Disrupting Compounds: Knowledge Gaps and Research Needs."
9. Williams, Richard *et al.* 2006. *Human Pharmaceuticals—Assessing the Impacts on Aquatic Ecosystems.* Setac Press.
10. Ternes, T., A. Joss, J. Kreuzinger, K. Miksch, J. Lema, U. VanGunten, C. McArdel, and H. Siegrist. 2005."Removal of Pharmaceutical and Personal Care Products: Results of Poseidon Project." *Proc. WEFTEC 2005,* Washington, D.C.
11. Oppenheimer, J., R. Stephenson and G. Loraine. 2005. "Characterizing Passage of Personal Care Products Through Wastewater Treatment Processes." *Proceedings of the 78th Annual Water Environment Federation Technical Exhibition and Conference,* Washington, D.C. Oct. 29–Nov. 2, 2005, Water Environment Federation: Alexandria, Virginia.
12. Oppenheimer, J. and R. Stephenson. 2006. "Characterizing the Passage of Personal Care Products through Wastewater Treatment Processor." *Proceedings of the 79th Annual Water Environment Federation Technical Exhibition and Conference.* October 21–25, 2006, Dallas, Texas.
13. Esposito, K. *et al.* 2007. "Multiple Barrier Treatment for Indirect Potable Reuse: Considerations for Trace Contaminants of Emerging Concern." WEFTEC 2007, San Diego, CA.
14. EPA. 2007. Draft Sampling Episode Report (March 2007).
15. Johnson, A. and R. Darton. 2003. "Removing Oestrogenic Compounds from Sewage Effluent." *Chemical Engineer,* 741.

16. Ying, G., R. Kookana and Y. Ru. 2002."Occurrence and Fate of Hormone Steroids in the Environment." *Environment International, 28*, 545.
17. Joss, A., H. Andersen, T. Ternes, P. Richle and H. Siegrist. 2004."Removal of Estrogens in Municipal Wastewater Treatment Under Aerobic and Anaerobic Conditions; Consequences for Plant Optimization." *Environmental Science & Technology, 38* (11), 0347.
18. Phillips, P. J., B. Stinsin, S.D. Zaugg, E.T. Furlong, D.W. Kolpin, K.M. Esposito, B. Bodniewicz, R. Pape and J. Anderson, 2005. "A Multi-Disciplinary Approach to the Removal of Emerging Contaminants in Municipal Wastewater Plants in New York State, 2003–2004." Proceedings of the 78th Annual Water Environment Federation Technical Exhibition and Conference [CD-ROM]; Washington, D.C., Oct. 29–Nov. 2; Water Environment Federation: Alexandria, Virginia.
19. Siegrist, H., A. Joss, T. Ternes and J. Oehlmann. 2005. "Fate of EDCs in Wastewater Treatment and EU Perspective on EDC Regulation." *Proc. WEFTEC 2005,* Washington, D.C.
20. Mueller, James A., *et al.* 1995. "Fate of Octamethylcyclo9tetrasiloxane (OMCTS) in the Atmosphere and in Sewage Treatment Plants as an Estimation of Aquatic Exposure." *Environmental Toxicology and Chemistry 14*(10) 1657–1666.
21. Ried, A., J. Mielcke. 2003. "Municipal Wastewater Treatment for Reuse: Combining Ozone and UV Techniques for Advanced Wastewater Treatment. Disinfection and Degradation of Endocrine Substances." *IUVA News, 5* (1), 17–21.
22. Helmig, Edward G., Rominder Suri, Patrick J. Cyr and Mohan S. Nayak. 2009. "API Removal from Pharmaceutical Manufacturing Wastewater: Results of Full-Scale Wastewater Treatment Performance." *Proc. WEFTEC 2009.*
23. Cleary, J.G., G.M. Grey, E. Helmig and R. Suri. 2011. "Treatment of Pharmaceuticals, Personal Care Products and Other Microconstituents—What Technologies are Working?" Workshop at WEFTEC, 2011, Los Angeles, CA.
24. PILLS (Pharmaceutical Input and Elimination from Local Sources), A European Partnership project of: Emschergenossenschaft (Germany), Waterschap Groot Salland (Netherlands), Centre de Recherche Public Henri Tudor (Luxembourg), Eawag (Switzerland), Glasgow Caledonian University (United Kingdom/Scotland) and Universite de Limoges (France). November 2010. "Pharmaceutical Residues in the Aquatic System—A Challenge for the Future."
25. Raczko, Bob, P.E. *et al.* 2013. "Evaluation of EDC Removal through WTP Processes." NJWEA 2013 Winter Technology Transfer Conferences, March 4–7, 2013.
26. Melcer, Henyk, Hugh Montieth, Charles Stapler and Gary Klecka. 2006. "The Removal of Alkylphenol Ethoxylate Surfactants in Activated Sludge Systems" *WEFTEC 2006.*
27. Johnson, A.C., H.-R. Aerni, A. Gerritsen, M. Gilbert, W. Giger, K Hylland, M. Jurgens, T. Nakarni, A. Pickering, M.J.-F. Suter, A. Svenson, F.E. Wettstein. 2005. "Comparing Steroid Estrogen and Nonylphenol Content Across a Range of European Sewage Plants with Different Treatment and Management Practices." *Water Res. 39*, 47–58.
28. Martz, Manfred, "Effective Wastewater Treatment in the Pharmaceutical Industry." Pharmaceutical Engineering (November/December 2012).
29. Salveson, Andrew, P.E. *et al.* "Trace Organic Compound Indicates Removal During Conventional Wastewater Treatment." WERF Final Report CEC4R08 (2012).

CHAPTER 7

Treatability Studies, Process Modeling and Troubleshooting

THIS chapter presents two key tools used today in combination for both design and troubleshooting of activated sludge and related processes. The first tool is treatability studies, which have been used for many years primarily for industrial wastewater but also for municipal wastewater projects with a high industrial wastewater contribution. Treatability study tools discussed herein include the following:

- Batch tests
- Feed batch reactor tests
- Respirometry testing
- Continuous flow bench scale tests
- Continuous flow pilot scale tests

Process modeling is the second tool being used more now in conjunction with treatability testing to develop key wastewater characteristics and kinetic parameters for model calibration. Modeling tools such as BIOWIN, GPSX, TOXCHEM and others have evolved over the last decade to become useful tools for process development, design and troubleshooting. Both of these tools are typically used now in combination on industrial wastewater projects to better understand the fate and removal mechanisms of specific organics including microconstituents in the activated sludge process. The modeling tools are also used for design and troubleshooting nitrification and denitrification at treatment plants as illustrated in the examples herein.

Troubleshooting of activated sludge problems including the loss of nitrification are presented along with a case study example. Design examples are presented to illustrate the use of treatability studies and process modeling.

TREATABILITY STUDIES

Treatability studies should always be used for design of industrial wastewater treatment projects. They are typically not needed for municipal wastewater treatment projects since the default values built into the process models are based on municipal wastewater characterizes and kinetic coefficients. The reason for use of treatability studies on industrial projects is that each industrial wastewater is unique in its range of wastewater characteristics relative to the following:

- Biodegradable and non-biodegradable fraction
- Presence of toxic and inhibiting compounds
- Inorganic salts and toxic metals
- Hydrocarbon and fatty acid based oil and grease
- Variety of particulates
- Clean-in-place cleaning chemicals
- Chemical compounds contributing to the COD
- Shock levels to activated sludge
- Chemical spills and shock loads
- Nutrient deficiency including micronutrients

The biodegradability or oxidation rates for industrial wastewaters varies with the type of industries and also with production changes and product campaigns.

Treatability testing is typically done by experienced personnel at laboratories, universities, technology vendors and engineering firms. Each test program is unique for the goals and objectives of the project requiring a specific test protocol. The tests are done to provide more confidence and reduced risk to the process design engineer that the process will work in the full-scale project. This chapter presents the types of treatability tests available plus some guidelines and examples of selecting the best type of test for the project. Other references [1] present a description of the batch and other biological treatment options available and a listing of the equipment needed.

Batch and continuous flow treatability tests have been used for years to develop process design criteria for industrial wastewater treatment plant design. The continuous flow tests are performed to simulate the actual or projected COD, BOD and nitrogen loading rates and sludge ages and determine the expected effluent quality. These tests are typically performed at a minimum of 3 sludge ages after an initial sludge acclimation period or typically a period of 2 to 3 months to obtain performance data at steady state conditions. These tests can be performed in the laboratory at bench-scale or as a pilot study at the plant. The pilot study has the advantages of developing performance data on day-to-day influent wastewater variability thereby providing more representation of a full-scale plant.

Techniques can be used in analyzing the less costly batch study data to develop treatment plant design criteria without the need for additional continuous flow and pilot testing. A methodology is presented [2] to conduct the batch tests and then develop continuous test *K* rate. The benefit of this approach is that it can save significant time and expense on a design project for an activated sludge treatment process. The *K* rate developed can be used to design completely mixed activated sludge processes including membrane bioreactors. A real case study is presented in this Chapter to show the difference in design for a pharmaceutical wastewater treatment nitrification system using batch test data versus continuous flow test data.

COD Removal Kinetics

The COD or BOD biodegradation rate (*K*) developed from the continuous flow bench or pilot tests is not the same *K* rate developed from the batch tests as follows:

$$\frac{S_o - S_e}{X_{vt}} = K(S_o/S_e) \quad \text{(Complete Mix Continuous Flow Test)} \qquad (7.1)$$

$$\frac{S_e}{S_o} = e^{-kX_v t} \quad \text{(Batch Test)} \qquad (7.2)$$

where,

S_o = Influent substrate concentration (mg/L)
S_e = Effluent substrate concentration (mg/L)
K = Kinetic coefficient (1/day)
X_v = Biomass under aeration (mg VSS/L)
t = Hydraulic retention time (days)

An approach can be used to develop the COD removal rate for a continuous flow test from the batch test data [1].

Wastewater characterization discussed in Chapter 2 and preparation of the wastewater for activated sludge treatability testing should be based on the equalized and pre-treated wastewater to simulate actual expected treatment plant operations. Depending on the nature of the wastewater and the discharge permit requirements, the following parameters should be included:

- BOD and/or COD or TOC
- total and volatile suspended solids
- oil and grease
- priority pollutants (VOC, SVOC)

- toxicity (bioassay)
- nitrogen forms (TKN, NH_3, NO_2^-, NO_3^-)
- phosphorus forms (ortho-PO_4, total P)

For wastewaters that do not contain aquatic toxicity, the following step-by-step procedure is applicable for all of the treatability study approaches:

1. Adjust the BOD:N:P ratio to 100:5:1, neglecting the wastewater organic nitrogen. Although organic nitrogen may be hydrolyzed to ammonia in the activated sludge process, it is initially neglected in order to insure adequate nutrients in the experimental phase. The availability of the organic nitrogen will be reevaluated in the final process design.
2. Evaluate the wastewater's potential to promote filamentous bulking. This can usually be evaluated by operation of a complete mix reactor at $F/M \approx$ 0.4/day for 5 to 8 days to establish the proliferation of filaments. Settling tests and microscopic examinations are used to characterize filamentous growth and affects.
3. Develop an acclimated mixed liquor. Determine the bioinhibition potential using the FBR procedure. For a wastewater with a low bulking potential, use a complete mix reactor. For a wastewater with a high bulking potential, acclimate the mixed liquor in a batch reactor, a sequencing batch reactor, or a biological selector. Adjust the initial feed rate of the wastewater to less than 50 percent of the inhibition threshold concentration. As acclimation proceeds, gradually increase the feed rate until the full waste strength is being treated.

Batch Tests

Batch tests which require much less time and cost are done by adding wastewater to an acclimated biological sludge for a period of 1 to 2 days. Batch studies have traditionally been used to characterize the biodegradability of a waste and develop COD and ammonia removal kinetic rate constants. The batch tests are relatively quick and inexpensive tests to perform compared to continuous flow tests in the laboratory and pilot tests in the field.

An example of batch test results for nitrification of a pharmaceutical wastewater are shown in Figure 7.1.

The resulting nitrification rates have an excellent correlation when presented as either NH_3-N removal or $NO_2 + NO_3$-N production (0.332 to 0.327 day^{-1}, respectively), based on the similar slopes of the ammonia removal and nitrite-nitrate produced. The nitrate production line is displaced by the initial nitrate concentration of 140 mg/L in the treated effluent used as the base substrate for the test. The nitrification rate developed from a second batch test was 0.366 at 20°C which agreed well with the first batch test. Multiple batch

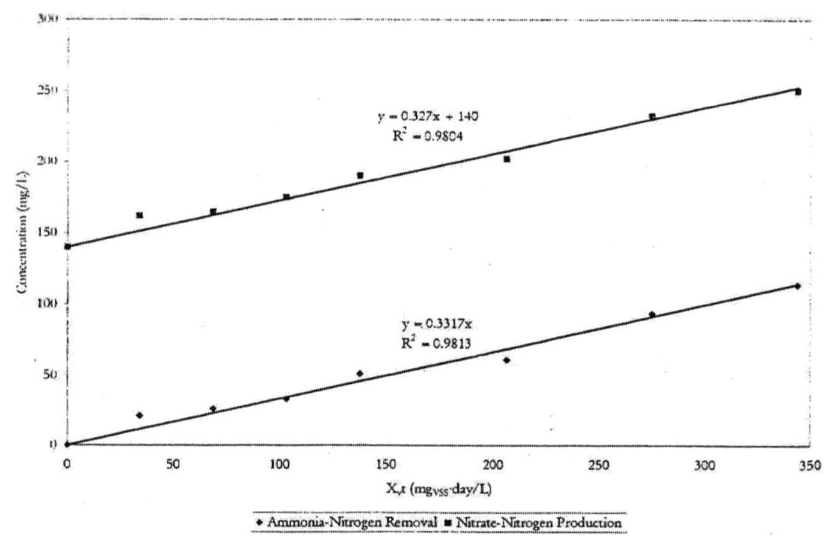

Figure 7.1. Ammonia-nitrogen removal and nitrate-nitrogen production observed during BAS testing.

tests should be done on different samples to see if results are consistent and reproducible. For wastewaters containing organic nitrogen, TKN should also be monitored to understand the transformation of organic nitrogen to ammonia and its influence on nitrate and nitrite nitrogen production.

Example 7.1—SBR Design for Pharmaceutical Wastewater Using Batch Tests

The following cxample shows how batch treatability tests results were used to develop sizing of a full-scale pharmaceutical wastewater plant to meet an acetone design basis limit of 2.1 mg/L. This project involved developing a wastewater solution to help the client treat its wastewater and meet the permit limits for discharge to the local Publically Owned Treatment Works (POTW). An alternatives evaluation was conducted to evaluate biological treatment alternatives with focus on achieving the acetone limit of 2.1 mg/L. The plant did not use acetone and suspected that it was a breakdown product from isopropyl alcohol (IPA) which was used as a cleaning solution in the production facility. A Sequencing Batch Reactor Solution was selected for design. An IPA recovery and source reduction program was also implemented to reduce the IPA loading and size of the SBR system.

Batch treatability tests were performed on the wastewater to refine and confirm process design parameters and performance estimates for the Sequencing Batch Reactor (SBR) plant design. Multiple tests were conducted to evaluate variability in wastewater characteristics and consistency in meeting the acetone permit limit. The tests were performed on actual wastewater samples collected from the site. These tests were done on representative samples to determine the variability in process parameters for evaluating biological treatment and the following goals and objectives:

- Determine the variability in the removal of COD, acetone and IPA from wastewater by conducting batch tests at food to microorganism loadings (*F/M*) that are approximately equal to, less than, and greater than the design conditions.
- Determine the biodegradation rates for both isopropyl alcohol (IPA) and acetone and the rate at which IPA will convert to acetone in a biological treatment plant.
- Refine and confirm the process design criteria for Sequencing Batch Reactor (SBR) aeration basin sizing to meet the acetone permit limit.
- Evaluate variability on wastewater characteristics and consistence in performance by conducting batch tests at loadings that are approximately equal to, less than, and greater than the design conditions.
- Evaluate sludge settleability, sludge production and potential for foaming.

The approach taken to design the SBR was as follows:

1. Obtain representative samples of the wastewater for characterization and design basis,
2. Perform bench-scale treatability studies in the laboratory to obtain biodegradation rates for acetone.
3. Scale-up biodegradation rate for full-scale SBR sizing.
4. Size full-scale SBR.
5. Compare full-scale SBR actual performance versus that predicted from the bench-scale tests.

A wastewater characterization sampling program was conducted at the plant. The results for the composite wastewater used for the study are provided in Table 7.1. A flow weighted composite wastewater sample from three sampling manholes had acetone and IPA concentrations less than the average design conditions. As a result, the feed solutions used for the batch testing were spiked with reagent grade acetone and IPA. In addition, the characterization results indicated the wastewater was nutrient deficient. Nutrients were added to satisfy the BOD_5: N:P ratio of 100:5:1 using ammonium chloride and potassium phosphate dibasic.

Five biological batch reactors and one volatile control reactor were set up with a combination of composite wastewater and unacclimated activated

TABLE 7.1. Composite Wastewater Characterization Results.

Parameter	Result
COD (mg/L)	1,022
Acetone (mg/L)	9.221
Isopropyl Alcohol (mg/L)	28
Temperature (°C)	18.9
pH (s.u.)	7.9
Alkalinity (mg/L as $CaCO_3$)	160
NH_3-N (mg/L)	2.13
Orthophosphate (mg/L)	2.52
Total Zinc (mg/L)	0.162
Dissolved Zinc (mg/L)	0.044

sludge and final effluent from a local municipal WWTP. Reactor 1 was a control reactor supplemented with dextrose for COD. Reactor 2 (design *F/M*), Reactor 3 (low *F/M*), and Reactor 4 (high *F/M*) were prepared with the 20 mg/L acetone and 405 mg/L IPA feed solution. The feed solution used for Reactors 5 (Spike) and 6 contained 50 mg/L acetone and 405 mg/L IPA. Reactor 6 was a volatile control reactor and did not contain any biomass.

The following results and conclusions were obtained from the batch tests:

- The COD data showed steady exponential removal with time in all reactors as shown in Figure 7.2. The control reactor substantially oxidized the degradable COD within 2 to 3 hours; the low *F/M* reactor at 8 hrs; the design *F/M* just after 8 hrs and the high *F/M* and Spike reactors between 8 and 21.5 hrs. All five biological reactors had similar final COD concentrations ranging between 42 and 48 mg/L, which represent the refractory organics that could not be further broken down biologically.
- The COD removals ranged from 82% in the low *F/M* reactor to 91% in the high *F/M* reactor.

Table 7.2 presents a summary of the IPA and acetone sampling data. The high initial acetone concentrations and low IPA concentrations at $t = 0$ suggest that the IPA immediately converted to acetone. After 1 hour, the IPA concentrations were non-detect (less than 13 mg/L) and remained non-detect throughout the batch test. The acetone data for Reactors 2 and 5 showed steady exponential removal with time. Similar acetone removal rates were observed in the design and spike reactor. After 4 hours the acetone removals in the design and spike reactors were similar at 70 and 66%, respectively. After 8 hours, the acetone removals were approximately 88 and 90% for the respective reac-

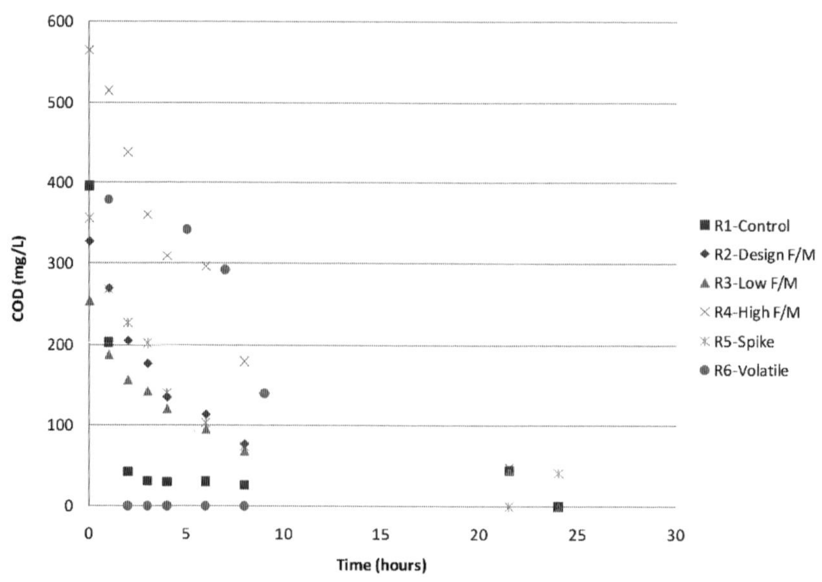

Figure 7.2. COD concentration vs. time.

tors. The acetone concentrations were reduced to non-detect levels (less than 0.0021 mg/L) in all biological reactors between 8 and 24 hours of operation.

Kinetics and Process Design Criteria

In order to confirm the aeration time required to meet the acetone limit of 2.1 mg/L, the reaction rate coefficient (K) was determined based on the following kinetic relationship for a batch system:

$$S_e = S_o e^{-kX_{vt/So}} \quad (7.3)$$

where,

S_e = Effluent acetone concentration
S_o = Influent acetone concentration
X_v = Mixed liquor volatile suspended solids (MLVSS0
T = Retention time
K = First order reaction rate coefficient

Based on the acetone data in Table 7.2, a plot of $\ln(S_e/S_o)$ versus $X_{vt/So}$ was prepared to determine the slope (K) for both the design and spike reactors as shown in Figure 7.4. The reaction rate coefficients ranged from 0.6445 to 0.725/day. In order to correct from bench to full-scale a factor of 2 was applied

TABLE 7.2. Summary of Acetone and IPA Concentrations.

	R2 (Design F/M)		R3 (Low F/M)		R4 (High F/M)		R5 (Spike)		R6 (Volatile)	
	Acetone	IPA	Acetone	IPA	Acetone	IPA	Acetone	IPA	Acetone	IPA
Feed Solution Target Concentration (mg/L)	20	400	20	400	20	400	50–75	400	50–75	400
Time (hrs)										
0	219						236			
1		**< 13**						**< 13**		
2	179						106			
4	65.7	**< 13**			142		80	**< 13**		
8	26.6						22.1			
22	**0.0021**	**< 13**	**0.0021**	< 13	**0.0021**	**< 13**				
24							**0.0021**	**< 13**		**< 13**

Bold values indicate sample was measured less than detection limit. Value shown represent detection limit.

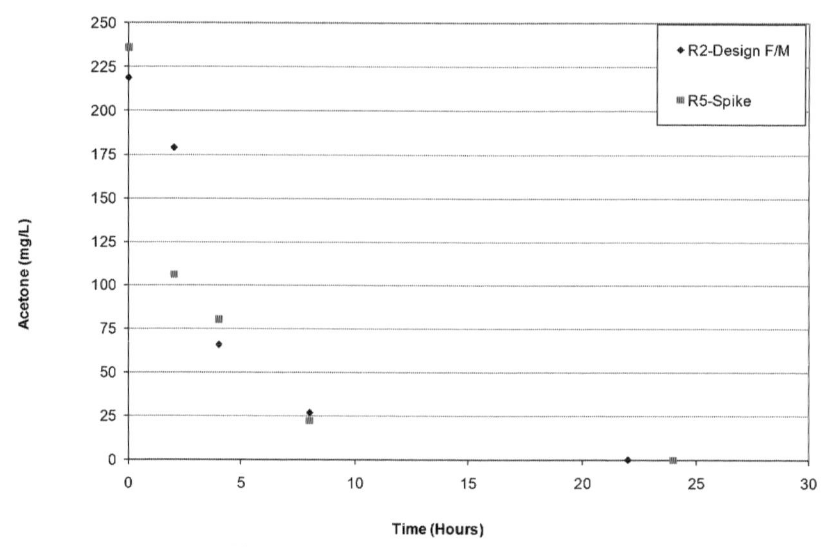

Figure 7.3. Acetone concentration vs. time.

(Eckenfelder *et al.*, 2010) and a coefficient of 0.35/d was selected for design. This reference suggests a scale up factor of over 3 but we recommended a factor of 2 since the batch test sludge from the municipal plant may not have been fully acclimated to the wastewater and that a higher K would most likely been achieved in a continuous flow test which allows sufficient time to achieve development of an acclimated sludge over a few weeks. This selected rate coefficient was similar to values from a literature review on acetone kinetics.

The design $K = 0.35/d$ was used to determine the aeration time required to reduce the acetone concentration in the full-scale SBR system to 2.1 mg/L. The initial SBR design prior to the treatability tests consisted of two (2) 13,629

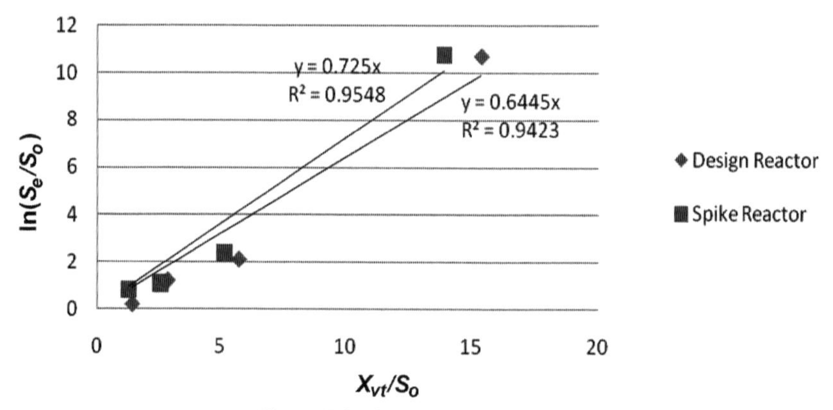

Figure 7.4. First order batch kinetics.

gallon reactors with each reactor operating at 4 cycles per day, 6 hours per cycle and a total aeration cycle of 4 hours. Based on dilution, the acetone concentration in the SBR reactors at average and peak concentrations would be 55 and 86 mg/L, respectively. At these initial concentrations and K = 0.35/*d*, the aeration time required to reduce the acetone to 2.1 mg/L at a design MLVSS of 3760 mg/L would range from 3.2 to 5.8 hours. This design consisting of a 4 hour aeration cycle would not be able to reduce the acetone at the peak conditions.

The Design Example 7.2 below was based on a reduction in IPA and acetone that would be required to meet the acetone limit of 2.1 mg/L (permit limit 3.0 mg/L) using the initial SBR design tank volume of 27,258 gallons operating at 4 cycles per day, 6 hours per cycle and a total aeration cycle of 4 hours. Based on dilution, if a 46% reduction in IPA and acetone can be achieved, the acetone concentration in the SBR reactors at average and peak concentrations would be 41 and 64 mg/L, respectively. At these initial concentrations and $K = 0.35/d$, the aeration time required to reduce the acetone to 2.1 mg/L at a design MLVSS of 3760 mg/L would range from 2.3 to 4 hours and the peak load conditions would satisfy the 4 hour aeration time requirement. The full-scale SBR pilot is performing better than expected which means the biodegradation rate may be higher than 0.35/day.

Design Example 7.1—Source Reduction IPA/Acetone Load and SBR Initial Tank Design (4 cycles/day, 6 hours per cycle)

Design Flow	20,500 gpd
Reactor Volume	13,629 gal, each at avg flow
Total SBR Volume	27,258 gal

6 Hours/Cycle (4× cycles per day for 2 reactors)

Gallons Treated	2,563 gal
Dilution Factor	0.19

Acetone Concentration
In Reactor Based on Dilution
(If initial Acetone/IPA = 455 mg/L) Co = 64 mg/L
(If initial Acetone/IA =218 mg/L) Co = 41 mg/L

$$S_e = S_o e^{-(KX_v/S_o)}$$

S_e = Effluent acetone, mg/L
S_o = Influent acetone, mg/L
X_v = MLVSS
T = Retention time
K = First order K rate (l/day)

Time Required for 2.1 mg/L Effluent Acetone (at peak)
0.166 days, 4.0 hrs, For $K = 0.35/d$

Time Required for 2.1 mg/L Effluent Acetone (at avg)
0.093 days, 2.3 hrs, For $K = 0.35/d$

Fed-Batch Reactor (FBR)

Fed-batch reactor tests have been used to determine nitrification kinetics [3] and removal kinetics of specific pollutants in activated sludge for years [4]. The essential characteristics of the FBR procedure are that: (1) substrate is continuously introduced at a sufficiently high concentration and low flow rate so that the reactor volume is not significantly changed during the test; (2) the feed rate exceeds the maximum substrate utilization rate; (3) the test duration is short and therefore allows simple modeling of biological solids growth; and (4) acclimated activated sludges are used.

A schematic diagram of the FBR is shown in Figure 7.5. Two liters of mixed liquor are placed in the reactor, and a sample is taken for determination of oxygen utilization rate (OUR) and mixed liquor volatile and total suspended solids prior to the start of the feed flow. The feed is introduced at a flow rate that exceeds the anticipated biodegradation rate and aliquots of the reactor contents are withdrawn every 20 min for the duration of the 3-hr test for analy-

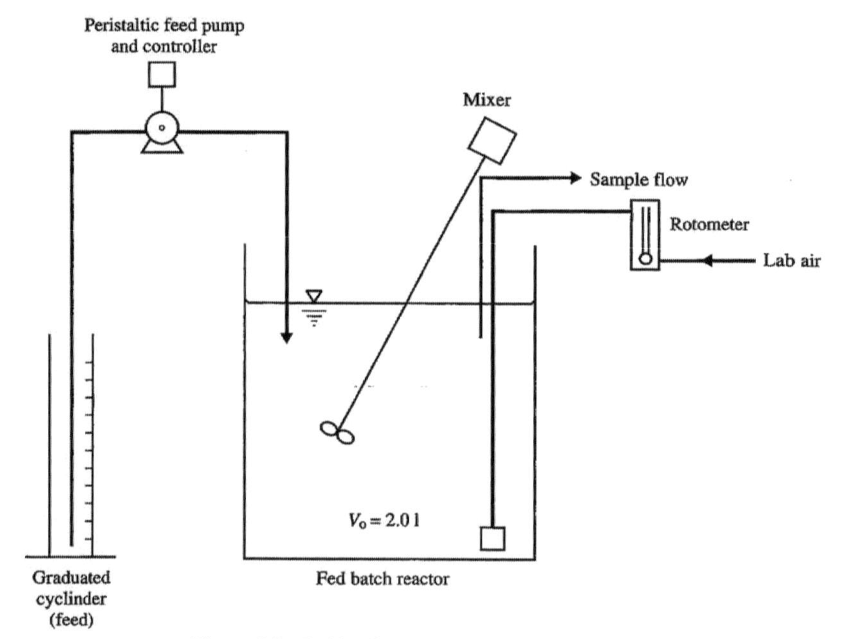

Figure 7.5. Fed batch reactor (FBR) configuration [5].

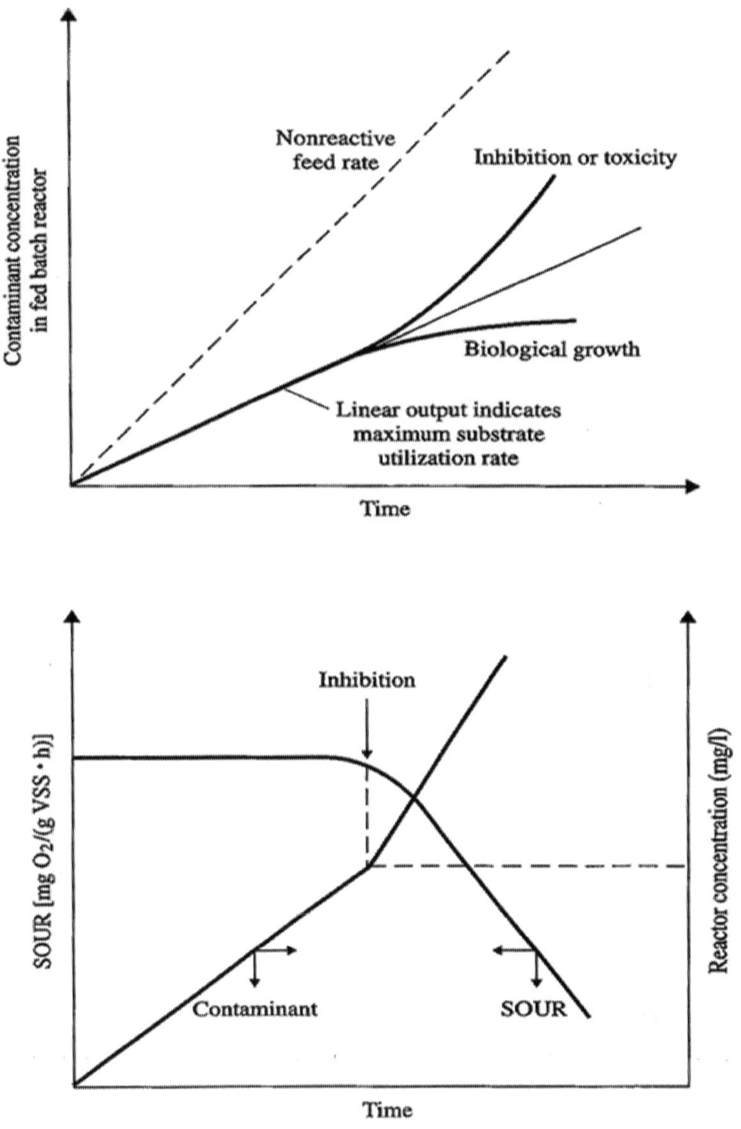

Figure 7.6. Theoretical fed batch reactor output with influent substrate mass flow rate > $q_{max} \cdot X_v$ and inhibition effects [4].

sis of the parameters of interest (COD, TKN, NH_5-N, NO_2 + NO_3-N). The OUR is determined in-situ every 30 to 60 min during the test. Suspended solids determinations are made every hour or two during the test.

The theoretical responses in a fed-batch reactor to both inhibitory and non-inhibitory substrates are depicted in Figure 7.6. In the case where substrate is added at a sufficiently high mass and low volumetric flow rate, the maximum substrate utilization rate will be exceeded, and the change in reactor volume will be insignificant. If the FBR volume change is negligible and the mass feed rate exceeds the maximum substrate utilization rate, then a substrate concentration buildup will result in the reactor with time. Noninhibitory substrate response results in a linear residual substrate buildup in the reactor with time. The maximum specific substrate utilization rate (q_{nax}) is calculated as the difference in slopes between the substrate feed rate and the residual substrate buildup rate divided by the biomass concentration. In the case of inhibition, substrate utilization would rapidly decrease, resulting in an upward deflection of the residual substrate concentration curve as shown in Figure 7.3. As inhibition progresses and acute biotoxicity occurs, the trace of the residual substrate concentration should become parallel to the substrate feed rate. The inhibition constant, K_I, can be approximated by identifying the inhibitor concentration at the midpoint of the curvilinear portion of the substrate response.

The modified FBR test described by Philbrook and Grady [3] is applicable to the determination of the kinetic coefficients q_m and K_S under field operating conditions. In the test, plant or pilot plant sludge at the desired SRT is placed in a 2-L reactor, and plant wastewater is added at a constant rate. In order to determine q_m, the addition rate must exceed the degradation rate. Since, in many wastewaters, the priority pollutant levels are low, the wastewater may have to be spiked to insure a sufficient concentration of pollutant to meet the conditions of the test. It is important, however, that the concentration levels achieved in the test are below the inhibition threshold. This can be found by the shape of the concentration-time curve. The degradation rate qm is computed as the difference in the slopes of the substrate addition rate and the residual substrate accumulation.

A second FBR test is then conducted with the addition rate of the pollutant of interest equal to one-half the maximum rate determined in the first test. The steady-state concentration observed in the reactor will be K_S. FBR test data for phenol are shown in Figure 7.7. Hoover [6] found a high variability in qm with sludges operating under the same loading conditions with time. Based on these observations, a routine test program should be established at a treatment plant and values for q_m and K_S interpreted on a statistical basis.

Example 7.2—Fed Batch Reactor (FBR) Test

A case study on a pharmaceutical industry wastewater study was used to compare using the batch and continuous flow treatability studies to meet project goals and objectives for ammonia-nitrogen removal. The FBR testing was conducted in accordance with the methods presented by Eckenfelder (Eckenfelder, Ford, Englande, 2009)[5]. Standard batch tests were also performed first for screening of wastewater samples.

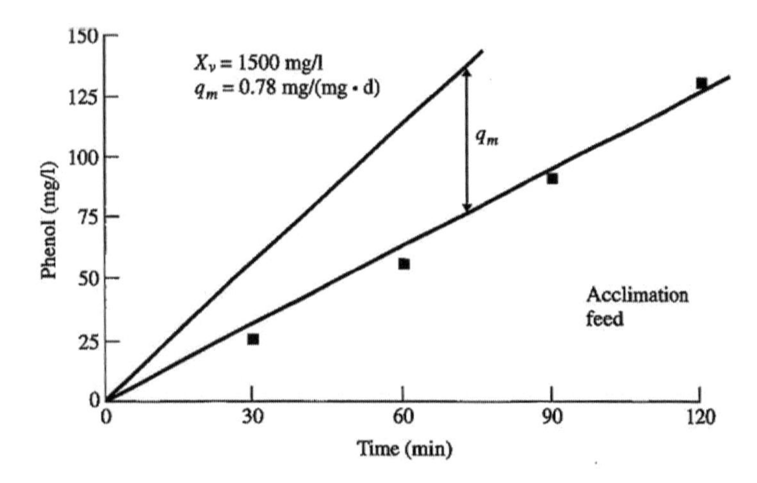

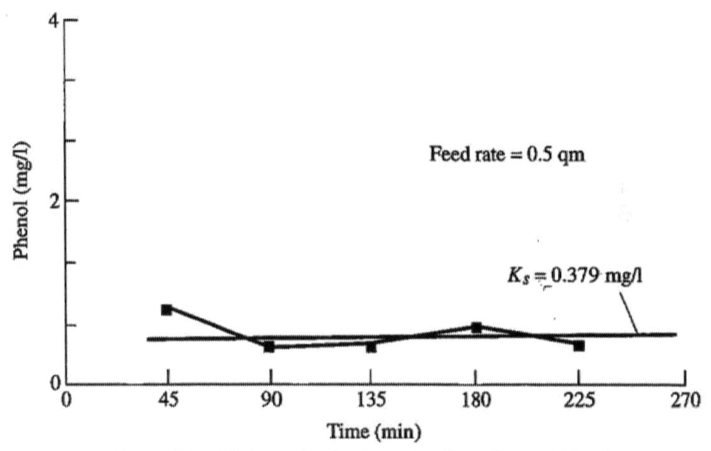

Figure 7.7. FBR test for the determination of q_m and K_s [4].

COD removal rates were calculated using complete mix kinetics as presented below:

$$(S_o - S_e)/(f_a X_v t) = K(S_o/S_e) \tag{7.4}$$

where,

S_o = influent COD
S_e = effluent COD
f_a = active fraction of biomass
t = hydraulic residence time
K = complete mix reaction rate coefficient

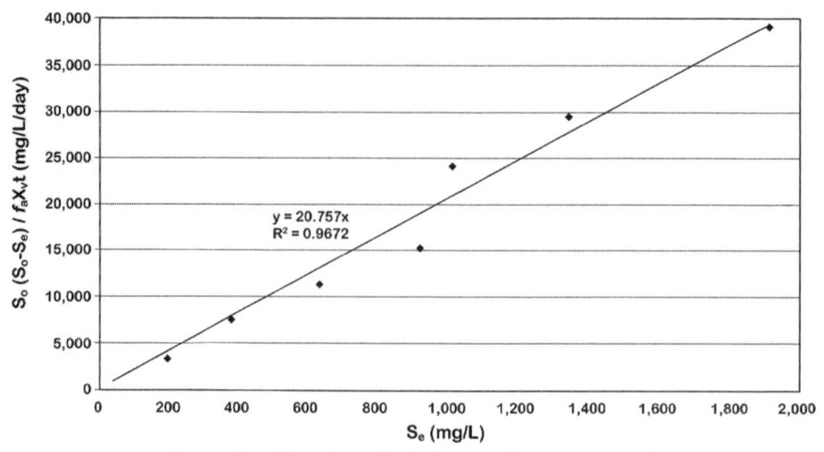

Figure 7.8. Plot of aerobic FBR data with $K = 20.8$ day^{-1}.

Manipulating the aforementioned equation into a $y = mx + b$ form allows the data to be plotted with the slope being equal to K. Figures 7.8 and 7.9 present the data plotted for the aerobic and anoxic tests, respectively.

Figure 7.8 shows the development of the COD removal rate from the FBR tests. The COD removal rate used for design of the aeration basin full-scale plant is 20.8 per day at 20°C. Figure 7.8 shows the COD removal rate developed from the anoxic reactor. The COD removal rate was 9.2 per day at 20°C. The calculated aerobic K of 20.8 day^{-1} on a COD basis can be converted to a BOD K of 14.3 per day. A typical BOD K for domestic sewage is 8 day^{-1} indicating the composite wastewater could be considered slightly easier to degrade than domestic sewage. The anoxic K of 9.2 per day can be converted to a

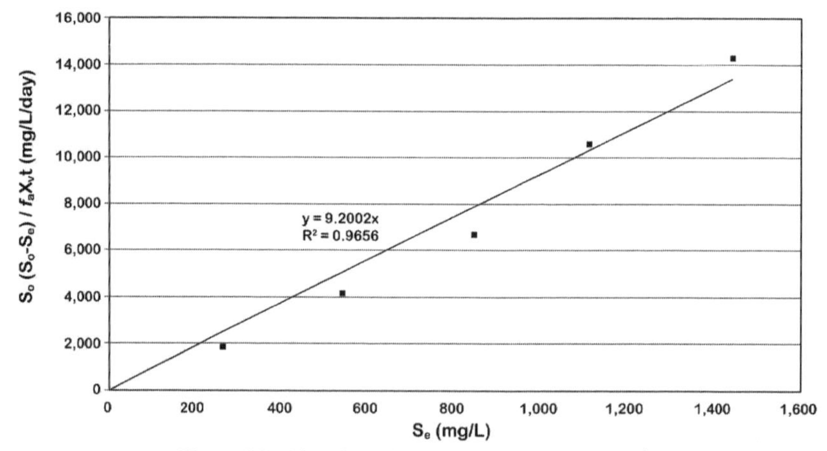

Figure 7.9. Plot of anoxic FBR data with $K = 9.2$ day^{-1}.

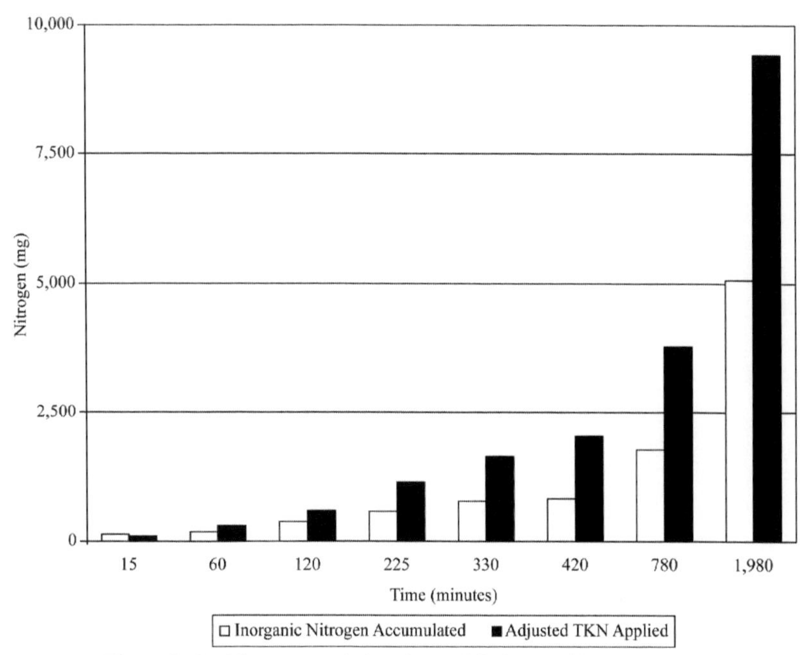

Figure 7.10. Nitrogen speciation observed during aerobic FBR testing.

BOD *K* of 6.3 day^{-1}. This rate is less than that for domestic sewage. The anoxic rate is directly related to the denitrification rate, as the COD is removed at the expense of nitrate. Based on stoichiometry, there are 3.43 mg O_2 available per mg NO_3-N. This compares well with the relationship observed in the testing, 3.11 mg COD consumed per mg NO_3-N reduced.

Ammonia Removal Rate

The aeration basin design volume in this project was driven by the ammonia-removal rate rather than the COD removal rate. Figure 7.10 presents the nitrification species data generated from the aerobic FBR test. The TKN applied in the composite wastewater less the amount of nitrogen required for cell synthesis is available for nitrification at 0.035 lb N/lb BOD removed and a degradable BOD/total BOD ratio of 0.68. The inorganic nitrogen accumulated represents the sum of NH_3-N and NO_2 + NO_3-N measured in the reactor at any given time. The difference between the adjusted TKN applied and the inorganic nitrogen accumulated is the TKN that did not hydrolyze and therefore is not available for biological removal or conversion. This fraction averages 0.49 of the adjusted TKN applied throughout the duration of the aerobic FBR test. The ammonia concentration expected following hydrolysis of TKN was approximately 2,500 mg/L.

Many industrial wastewaters do not inhibit carbonaceous oxidation but are inhibitory to the more sensitive nitrification process. Nitrification inhibition kinetics can be studied using the FBR procedure or a batch test protocol. Depending on the source of the toxicity, both test methods can include addition of powdered-activated carbon (PAC) to absorb the toxicant. Alternatively, the raw wastewater can be pretreated by chemical oxidation or precipitation to reduce the toxicity prior to the nitrification test procedure.

CONTINUOUS FLOW BENCH SCALE TESTS

Example 7.3—Continuous Flow Test

Continuous flow bench studies take longer then batch tests but provide a simulation of what would be expected in a full-scale plant operation including daily changes in organic and nutrient concentration and wasting of biomass to maintain a consistent sludge age. For the wastewater in Example 7.2, a continuous flow test was also conducted for three months to simulate the Modified Ludzack Ettinger (MLE) process for organic removal, nitrification and denitrification. The MLE process includes an anoxic reactor for nitrate and organics removal followed by an aeration basin for nitrification and residual organics removal. The aeration basin mixed liquor containing the high nitrates is recycled back to the anoxic reactor at a rate of 3 to 1 and higher. Batch tests were also performed to evaluate kinetics and potential for inhibition to the nitrifiers prior to the continuous tests. Figure 7.11 shows a picture of the laboratory reactors.

A nitrification rate range of 0.33–0.47 mg NH_3-N/mg MLVSS-day at 20°C was developed from the continuous treatability test data. If the batch tests were used to design the plant without doing the continuous treatability study tests, the same aeration basin volume would still have been used for design.

Table 7.3 shows a summary of the initial assumed conceptual design values prior to conducting any batch or continuous tests versus what we determined in the batch tests on simulated wastewater samples.

The assumed value of "*a*" is within normal experimental error of the determined value. By using the assumed value of "*a*", the conceptual level design is somewhat conservative by assuming the sludge production is higher than that determined experimentally. This conclusion is similar for the assumed value of "*a′*". Since the assumed value was higher than the experimental value, greater oxygen transfer capacity was provided than may be required in actual practice. The conceptual level design was conservative by providing additional oxygen transfer and associated aeration basin tankage. The BOD removal rates are also very similar. The assumed aerobic *K* is less than the experimentally determined value. The exact opposite is true for the anoxic *K*. The assumed value is

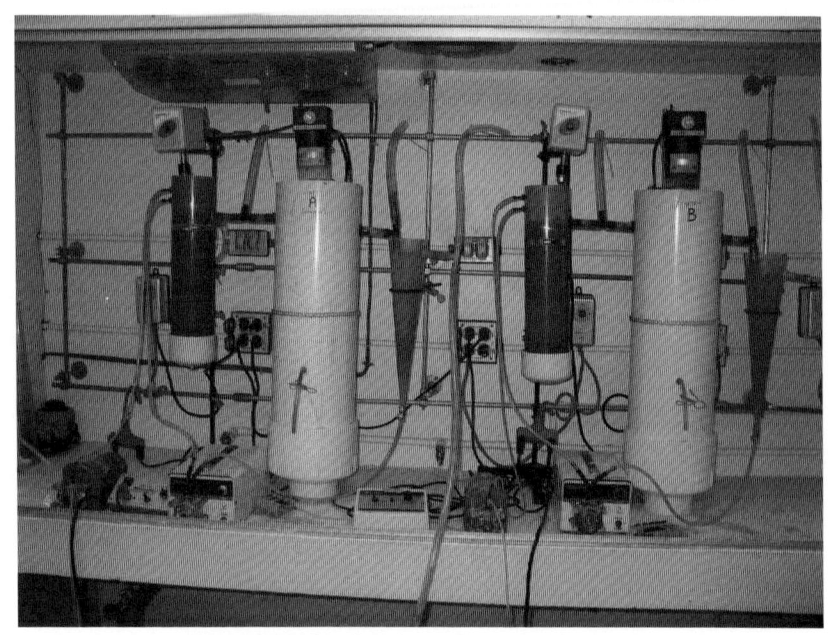

Figure 7.11. Bench-scale anoxic and aerobic reactors.

25 percent greater than the calculated value. The conceptual level designs were conservatively high. The nitrification rate used for the conceptual design was 0.35. This compares favorably with the calculated value of 0.33.

In summary, this example showed the following:

1. Batch treatability tests are the quickest and most cost-effective approach for obtaining process design kinetics for COD, BOD and nitrogen removal.
2. COD removal rates from batch tests can be used to develop COD kinetic removal rates for continuous tests for use in design.

TABLE 7.3. Summary of Assumed Values and Values Determined by Batch Treatability Testing.

Parameter	Assumed Conceptual Design Value	Determined Value	Units
a	0.6	0.56	mg VSS/mg BOD
a'	0.6	0.49	mg O_2/mg BOD
$K_{aerobic}$	12	14.3	day^{-1} on a BOD basis
K_{anoxic}	8	6.3	day^{-1} on a BOD basis
Nitrification Rate	0.35	0.33	mg NH_3-N/mg VSS day

3. For the pharmaceutical wastewater case study, the COD removal rate for the continuous flow tests were approximately 40% higher than the batch test rate for the aerobic reactor and 21% higher for the anoxic reactor.
4. The ammonia removal rate from the case study for the continuous flow tests of 0.366 mg $NH\text{-}N_r$/mg MLVSS-day agreed well with those from the batch tests at 0.33 to 0.47 mg $NH\text{-}N_r$/mg MLVSS-d_2.
5. The full-scale plant aeration basin design volume which was driven by the ammonia-removal rate from the continuous flow tests would have been about the same if it was designed from the batch test kinetic rates.
6. In hindsight, the batch testing would have been sufficient to develop the design of the full-scale plant aeration basins. The continuous tests did provide more confidence in the design by testing changing wastewater characteristics over a period of time.

CONTINUOUS FLOW PILOT SCALE TESTS

A continuous flow pilot study is the best treatability study design tool for a number of reasons. It is done at the plant and allows larger size aeration basins and clarifiers. It also allows evaluation of the impact of the continuous day-to-day of variability in influent flow and wastewater characteristics on the activated sludge performance including nitrification and gives the plant personnel some training in the operation and monitoring. The continuous bench scale tests typically change the influent feed samples less frequently at twice per week or less. Pilot plant procedures are described in other books and refineries [3,5,6]. The day-to-day sampling and monitoring tests are similar to continuous bench scale testing.

Two or three reactors are typically operated in parallel over an SRT range of 3 to 12 days for a readily degradable wastewater and 10 to 40 days for a less degradable wastewater. The operating temperature also affects the selection of SRT. The SRT should be maintained by daily wasting of an appropriate mixed liquor volume (i.e., for a 10-day SRT, one-tenth of the reactor volume is wasted daily). The waste sludge mass is computed as the VSS in the wasted volume, plus the VSS in the reactor effluent. A sampling and analytical schedule for the reactors should be developed with daily monitoring of pH, DO and oxygen uptake and with sampling of influent and effluent parameters 2 or 3 times per week. Some consideration should also be given to reactor configuration. Three or more reactors in series shall be used to simulate plug flow operations while a single reactor should be used to simulate complete mix operations.

Example 7.4—High TDS Wastewater—MBR Pilot Study

The challenge on this project was to treat a pharmaceutical plant wastewater for BOD removal with the wastewater containing 100,000 mg/L TDS. Activated sludge treatment is typically limited to about 20,000 to 30,000 mg/L TDS at which point defloculation of the biomass occurs and settleability deteriorates. With the availability of membranes in MBR systems, activated sludge can now be used to treat high TDS wastewaters since the membrane retains the biomass in the system. This pilot study demonstrated an average BOD removal of 77% at a range of BOD concentration from 275 average to 1,170 mg/L peak. The TDS concentration ranged from 80,000 to 120,000 mg/L.

RESPIROMETRY TESTS

Respirometry devices called respirometers are also used to measure oxygen utilization or uptake rate. There are cylindrical shaped reactors similar to those used for batch tests. The oxygen utilization is typically measured continuously and cumulatively to show overall oxygen consumption over time. On-line respirometers are also used at treatment plants to provide an early warning of changes.

PROCESS MODELING

The design of industrial wastewater treatment plants are wastewater specific at each facility. Batch and continuous flow bench-scale treatability studies and pilot plant studies are almost always performed in the design development phase of a project. Process modeling tools can be used to develop sizing of treatment facilities using the process design criteria such as COD and nitrogen removal kinetic constants developed from the treatability testing. Batch tests typically require less time and costs than continuous bench testing and pilot testing and can be used on a fast track project schedule with budget constraints. The batch tests together with process modeling tools may or may not result in the most optimal design and facility sizing versus the continuous and pilot scale tests which offer more certainty in testing the variability in wastewater characterization over several weeks or months where batch tests are typically done on a few samples over a shorter time frame.

Steady and dynamic modeling of wastewater treatment plants (WWTPs) offers valuable insights into plant design, operations, trouble-shooting, optimizing plant process operations, response to loading variability and training plant staff. The application of the commercially available activated sludge models (i.e., BIOWIN and GPSX) to the modeling of plants treating domestic wastewater has been well documented [6] and these models are configured with default valves representative of domestic wastewater. The simulation of industrial activated sludge plants poses special challenges due to the unique and highly variable nature of the wastewater and the range of operating conditions often encountered. There are special challenges associated with modeling industrial and municipal WWTPs. Model enhancements are often required to handle microconstituents including pharmaceuticals. Examples of pilot-testing, scale-up, design, lessons learned, and future modeling trends include:

- Tracking the dynamic fate of specific trace organic compounds (including metabolite formation and removal;
- Modeling the effect of inhibitory levels of substances on biological competence, contaminant fate and biomass settling characteristics; and
- On-line monitoring with potential to include warnings of imminent upsets and knowledge-based solutions.

Modeling tools such as BIOWIN and GPSX continue to be utilized more in the design and troubleshooting of wastewater treatment systems. The use of these models and the selection of the best treatment technologies require a good understanding of the basic principles of biological treatment for organics and nutrients removal. These models are now being in conjunction with other chemical fate models such as Water 9 and TOXCHEM to understand the fate and removal mechanism of specific chemicals or classes of chemicals such as microconstituents through POTWs and industrial wastewater plants. These tools consider chemical volatility and sorption along with biodegradation.

Removal of COD and BOD is still used in the design of industrial treatment plants, however, effluent limitation guidelines and permits also require design for removal of specific organic constituents such as benzene, toluene, phenol and other industrial categorical parameters. The modeling tools contain the basic process design equations and variables such as: biodegradation kinetics and rates, oxygen utilization coefficients, sludge yield, temperature and dissolved oxygen corrections, etc. which has not changed over the years. The available modeling tools also allow the user to perform time-variable simulations and sensitivity analysis to design parameters to help decision-makers select the best technology solution. The models now include the various biological treatment processes including: activated sludge (plug flow and completely mixed); SBRs; MBR; MBBRs; IFAS and others.

MODELS FOR FATE OF CONSTITUENTS THROUGH TREATMENT

Available process models such as TOXCHEM and Water 9 have been used to evaluate the fate and removal through treatment plants of specific chemicals or classes of chemicals using biodegradation rate coefficients, volatilization rates and partitioning coefficients. STPFATE9 is an in-house model developed by HydroQual [7] to determine the fate of organic contaminants through a wastewater treatment plant. These models evaluate the mechanisms of biological degradation, volatilization and adsorption for a typical activated sludge process with secondary clarification. The adsorption mechanism results in a portion of the contaminant of concern being wasted in the secondary sludges. The total concentration of the contaminant of concern is distributed between three phases, (1) the free dissolved phase, (2) the dissolved phase held in solution by dissolved organic carbon (e.g., humic type substances) and (3) the particulate phase sorbed onto the suspended solids. Volatilization occurs in the activated sludge aeration tank. The aeration is assumed to be forced diffused aeration. Steady state conditions are assumed in all processes. The model uses two physical chemical properties of the contaminant of concern to determine the fate of treatment. The octanol-water partition coefficient quantifies the degree of sorption and the Henry's Law constant conveys the extent of volatilization. The models also include a temperature correction for the Henry's Law constant. The biodegradation rate of the contaminant of concern is also required.

Some consultants, universities and professional organizations have developed their own chemical fate models. STPFATE was validated using available literature data from two pilot plant studies and two full-scale plants. Removal data were available for seven organics that had varying physical chemical properties such that each removal mechanism could be validated. From all analyses it was demonstrated that STPFATE could be used to accurately predict the concentrations of organics in the effluent and sludge. The ability of STPFATE to predict the fate of organic compared favorably with the results from five other treatment plant models. Tools like STPFATE are useful in analyzing chemical fate only if they area calibrated and validated.

The following approach has been used to evaluate and compare treatment alternatives for microconstituents including active pharmaceutical ingredients (APIs) and insecticides. This approach is based on understanding the physical/chemical characteristics of the microconstituents to be treated or controlled and their fate or removal through various treatment processes. This understanding can be developed through the use of the following approach and sequence of work tasks:

- Review of physical/chemical properties which impact a microconstituent's behavior through treatment processes (i.e., biodegradation, volatilization, adsorption, etc.);
- Benchmark comparison of treatment alternatives and costs through available literature;
- Use of modeling tools to predict the fate or removal through existing

and proposed treatment systems;
- Use of desktop alternative analysis to compare alternatives based on cost and non-cost criteria;
- Use of bench-sale treatability studies to further refine the desktop alternatives comparison and cost comparison;
- Use of pilot studies to further refine process design and expected treatment performance over a period of time and variability in loading conditions.

The physical and chemical properties are critical to using the process modeling tools to predict performance through treatment. The treatability study or full-scale sampling data can then be used to calibrate the models and subsequently used to predict performance with optimization of the treatment process (i.e., increased sludge age) or adding additional treatment capacity or processes.

The present emphasis on volatile emissions from wastewater treatment plants requires that stripping be considered in activated sludge process design where volatiles are present in the influent wastewater. There are several factors to be considered:

- Both the power level in the aeration basin and the type of aerator (i.e., diffused or mechanical) significantly influence stripping;
- The maximum expected concentration of each particular volatile should be employed. The degradation rate of specific volatiles will be related to both the composite wastewater composition and the process operating conditions, i.e., the SRT;
- It has been shown that off-gas capture and recirculation will significantly enhance biodegradation of VOCs. Therefore, this process modification should be included in the pilot studies if VOC emission control is required for the plant;
- Covered aeration basins may result in a significant temperature rise due to the exothermic reaction with high-strength wastewaters. Basin temperatures in excess of approximately 38°C may result in floc dispersion and inhibition of nitrification and other treatment processes. In these cases, it is necessary to monitor the reactor temperature and make appropriate temperature adjustments as required.

Example 7.5—Full-Scale Plant Process Model

A process model BIOWIN was used to simulate full-scale plant performance based upon current plant data. The pharmaceutical wastewater treatment plant operating data from September 1, 2008 through October 31, 2008 was used to calibrate the treatment plant model. A process flow diagram of the model of the existing treatment plant is shown in Figure 7.12.

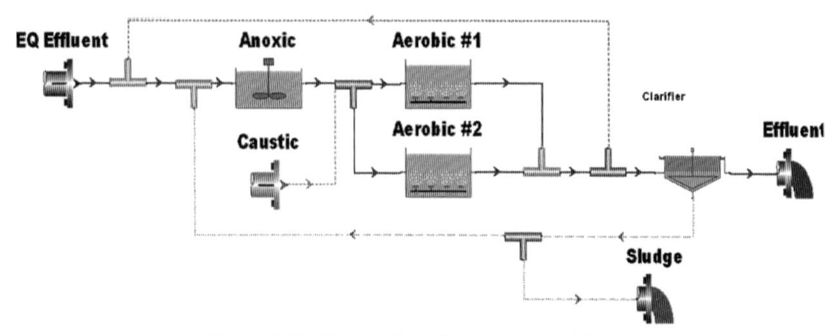

Figure 7.12. Process flow diagram for model [11].

The influent for simulation comes from a 100,000 gallon equalization tank which is monitored by the plant for chemical oxygen demand (COD), ammonia, pH and total suspended solids (TSS). The model was run at steady state conditions and did not include the equalization basin. The existing activated sludge plant includes the Modified Ludzack Ettinger (MLE) process for nitrification and denitrification using an anoxic selector. The plant operates at a MLSS of 5,300 mg/l with a SRT of 12 days. There is one secondary clarifier. Table 7.4 shows the actual plant performance based upon existing operating conditions and plant influent characteristics for two data periods compared to the model prediction. Good agreement was obtained with the model.

The calibrated model was then checked using peak full-scale plant data from September through October 2008 which represented a peak loading period. A dynamic simulation was done for the peak operating data and the model results can be seen in Figure 7.13.

Figure 7.14 shows the full-scale operating data from the plant from September and October 2008.

The breakthrough of effluent ammonia can be seen in both the model and the full-scale operating data for September through October 2008 indicating good model calibration to the nitrification kinetics.

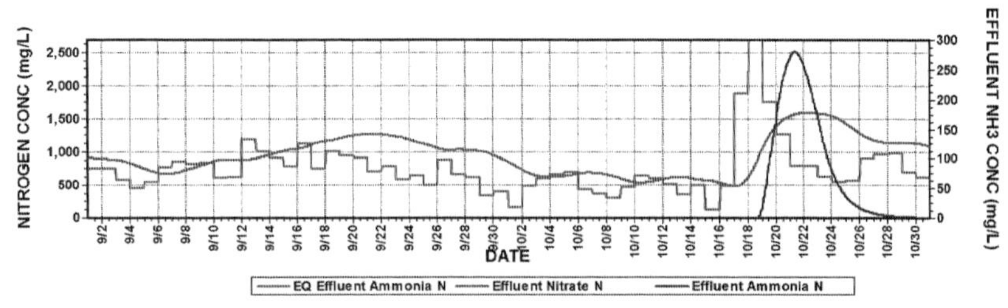

Figure 7.13. BioWin dynamic simulation of influent ammonia vs. effluent nitrate vs. effluent ammonia [11].

TABLE 7.4. Model Calibration Comparison [11].

Parameter	Condition 1		Condition 2	
	Actual Plant Performance	Model Prediction	Actual Plant Performance	Model Prediction
MLSS, mg/l	5,300	5,200	5,500	5,600
Sludge Wasting, lbs/day	3,100	2,672	3,000	2,870
Sludge Yielded, lbs VSS/day/lb COD removal	0.33	0.33	0.35	0.35
Oxygen Required, lbs/day	12,521	10,366	9,400–11,100	10,350
Effluent COD, mg/l	100	129	125	137
Effluent NH_3, mg/l	< 0.5	< 0.5	< 0.5	< 0.5
Effluent TKN, mg/l	–	11	–	16
Caustic Usage, gal/day	167	200	167	200

TROUBLESHOOTING

Troubleshooting of activated sludge plant problems such as bulking was previously discussed in Chapter 3. Chapter 3 showed examples of well developed and healthy activated sludge floc. Microscopic analysis is a valuable tool to assess the microbiology characteristics of the biomass in the activated sludge plant. This Chapter describes an example of using treatability testing and process modeling tools to troubleshoot, conduct root cause analysis and develop a recovery plan for nitrification upsets. The use of microscopic analysis to investigate activated sludge performance problems such as filamentous and viscous bulking are presented by others [8].

This information is based on actual problems and case studies at a variety of plants. A good overview of process variables impacting performance has been presented in Chapter 3. If plants are designed and operated based on the guidelines in this book, they should perform well. Performance issues are

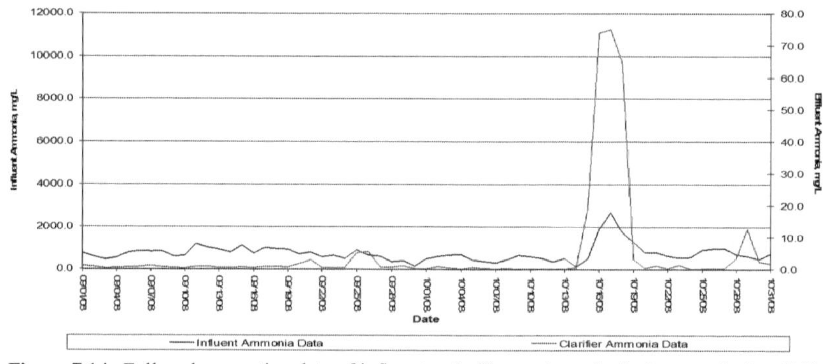

Figure 7.14. Full-scale operating data of influent and effluent chronological ammonia data [11].

typically related to a change in wastewater characteristics or a change in plant operations including the following [9].

Changes in Wastewater

- Higher or lower than normal flow, organic loading or nitrogen loading;
- Change in production to a new product and therefore change in wastewater characteristics;
- Change in Pre-treatment process removal.

Changes in Operation

- High or low sludge age;
- Low dissolved oxygen;
- Change in pH or wastewater buffering capacity;
- Change in nutrient addition;
- Shock load from a spill or production variability; and
- Addition of an inhibiting compound which impacts nitrification.

Analysis of historic operating data can offer help in the assessment of plant performance problems and troubleshooting. There are a number of troubleshooting tools that can be used to determine the activated sludge biomass condition including:

- Oxygen uptake rate
- Microscopic analysis
- Sludge volume index (SVI) and other settleability tests
- pH
- Dissolved oxygen
- Oxidation reduction potential (ORP)
- Active biomass testing such as ATP monitoring
- Treatability tests/respirometry

A combination of these tools are typically used for troubleshooting upsets. Additional sampling of the influent and effluent from the activated sludge process is also included in the troubleshooting to check on parameters such as nutrients and micronutrients. These tools are described in detail in other books [1]. The example of troubleshooting presented below utilized all of the above tools except for ATP monitoring which is utilized more now particularly in pulp and paper plants.

Example 7.6—Troubleshooting Loss of Nitrification

This case study involved the loss of nitrification in a pharmaceutical plant

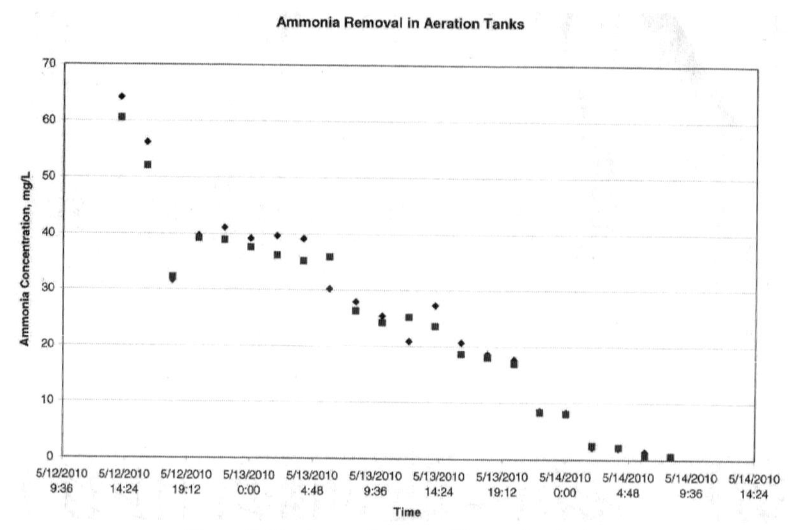

Figure 7.15. Ammonia removal in aeration basins.

which uses the MLE activated sludge process for nitrification and denitrification. The full-scale plant began to experience a reduction of its ability to remove ammonia which caused an exceedance of its permit limit and an increase in ammonia nitrogen in the aeration basin. The ammonia nitrogen concentration in the aeration basin increased to about 70 mg/L which resulted in an un-ionized ammonia concentration which was inhibiting. The plant's standard operating procedure (SOP) was immediately implemented and included adding two gallons of the nitrifier solution to the aeration basin plus receding with a nitrifying sludge from a local POTW. The ammonia decreased to less than 1 mg/L in 34 hours as shown by the data in Figure 7.15. The ammonia removal rate developed from this data was 0.16 to 0.31 mg NH_3-N_r/mg MLVSS-d at MLSS of 2,500 mg/L, 85% VSS and a temperature of 32°C. These data represent a full-scale batch test since the influent flow rate to the aeration basin was shut off. If the ammonia release from endogenous respiration of the sludge is taken into account, these rates could be about 60% higher at 0.25 to 0.50 mg NHr/mg MLVSS Nit^{-day} which agrees with the batch and continuous flow tests kinetic rates. A team was assembled by the client to develop a recovery plan for nitrification and determine the root cause of why it happened. Treatability testing and the existing plant process model were used to assist the team in the recovery and to identify potential root causes of loss of nitrification.

Recovery Plan

- Curtail plant production to minimize and reduce the ammonia feed to less than 500 pounds/day and add dextrose solution for COD.

- Add bicarbonate alkalinity to buffer the pH to around 7 to minimize inhibition due to unionized ammonia.
- Add nitrifying sludge from local POTW.
- Add 2 gallons of pure nitrification solution initially and then 1 gallon at a time every few days.
- Start daily sampling and analysis for COD, TKN, ammonia, nitrite and nitrate on the influent and effluent from the anoxic selector, aeration basins, plus influent from the equalization tanks.
- Begin batch treatability testing on wastewater source streams and night-time CIP cleaning streams.
- Reduce wasting to increase sludge age from 12 to 20 days.
- Add phosphorus and trace nutrients.
- Lower pH in aeration basin to 7 to reduce unionized ammonia.

The following most probable root causes were identified and evaluated:

- Heavy acid addition and drop in pH to 4 in one of the two aeration basins due to pH control system malfunction.
- Septic conditions and odor in the EQ tank.
- Copper in the wastewater.
- Higher than normal CIP cleaning chemical usage during night shift.

Both treatability testing and process modeling were used in the plant recovery. The treatability testing showed that all of the source wastewaters coming to the plant at the time of the upset except for CIP cleaning wastewater were not inhibitory to nitrifiers. The CIP was shown to be inhibitory at high loadings. The process model which was developed and calibrated previously and discussed in Example 7.4 was used to simulate the ramp-up of TKN load with time and estimate the recovery time. A recovery time of 10 days was estimated and used to guide the operators on what TKN loading to use each day as the plant was operated on a batch basis from the equalization tanks. The model continues to be used today to optimize how best to upgrade plant capacity to handle future production increases.

Example 7.7—Fate of Pyrethroids through Activated Sludge

A project was performed as part of the research in understanding the fate of pyrethroids through wastewater treatment processes. Based on physical-chemical properties, pyrethroids are not expected to be volatile and they should have high sorption to solids. Two studies were conducted where pyrethroids were monitored. One was a bench-scale study where plant influent was spiked with a mixture of pyrethroids and treated through conventional treatment processes. The other study was conducted at a POTW and involved full-scale sampling and modeling of pyrethroids through the plant. The main goals of these projects were as follows:

1. Determine the removal of pyrethroids through conventional treatment processes of primary settling, aerobic treatment, anaerobic digestion and ultrafiltration;
2. Determine the fate distribution of pyrethroids through wastewater treatment (e.g., removals via volatilization, sorption, biodegradation);

Four treatment processes were simulated in bench scale studies primary sedimentation, aerobic biological treatment, anaerobic biological treatment and ultrafiltration. Local POTW plant influent was spiked with a mixture of the pyrethroids to achieve target levels of 5 μg/L of each of pyrethroids except for permethrin which had a target level of 50 μg/L. The primary effluent was further treated via aerobic biological treatment in a continuous flow-through system. The system was operated at a target SRT of 10 days and a *F/M* of 0.3 g COD/g MLSS-d for 30 days. The effluent from the aerobic biological system was further treated via filtering to determine if any additional removal of the pyrethroids was achieved. Ultrafiltration was accomplished using a vacuum filtration apparatus with filter pore retention sizes of 0.1 and 1.0 μm. Based on mass balance computation around the aeration system, pyrethroid removals via biodegradation ranged from 34.2 to 81.6%. Filtering through either a 0.1 or 1.0 μm filter achieved greater than 90% removal of the residual pyrethroids.

An in-depth sampling and modeling effort was conducted at a POTW to understand the fate of the pyrethroids through the treatment process. The major treatment processes include primary settling, biological treatment with a pure oxygen supply, secondary settling, chlorination and de-chlorination, sludge thickening and anaerobic digestion. The work consisted of three tasks: compilation of operational and available monitoring data, full-scale sampling and analysis of pyrethroid concentrations, and process modeling of the facility. The modeling software used for this work was TOXCHEM, a commercially available program that predicts the fate of trace organic contaminants, through wastewater collection and treatment processes. TOXCHEM is calibrated based on flow and solids concentration data for the unit processes. Three sampling events occurred where solids, flow and pyrethroid data were obtained. The treatment train at a POTW was simulated in TOXCHEM. Once the model was calibrated, an initial run of the pyrethroids was performed and then the biodegradation rates and log KOW coefficients were calibrated by comparing predicted and measured concentrations in process samples collected from the sampling program.

Biological treatment and secondary clarification reduced the pyrethroid concentration by an additional 85 to 93%. On average the total removal of the individual pyrethroids through the plant ranged from 89 to 95%. The predicted distribution of the pyrethroids is shown in Figure 7.16. The median removal

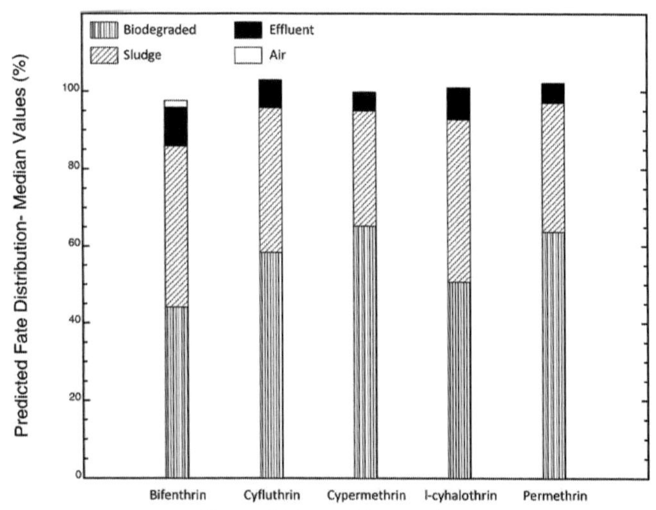

Figure 7.16. Modeling results.

for biodegradation ranged from 44 to 65%. The median removal for sorption ranged from 30 to 42%. Total pyrethroid discharged to the effluent ranged from 3 to 12%. With the exception of bifenthin, emission to the air was predicted to be negligible. Figure 7.16 shows the modeling results and the fate of five of the pyrethroids with a high percentage received by biodegradation and partitioning to sludge.

REFERENCES

1. *Industrial Wastewater Management, Treatment and Disposal,* Third Edition, Manual of Practice No. FD-3. Water Environmental Federation (WEF). 2008.
2. Cleary, Joseph G., W.W. Eckenfelder, G.M. Grey, R. Orlando. 2010. "Using Batch Activated Sludge Treatability Testing in Process Design." *WEFTEC 2010.*
3. Williamson, K.J. and P.L. McCarty. 1975. "Rapid Measurement of Monod Half-Velocity Coefficients for Bacterial Kinetics," *Biotechnology and Bioengineering, 17*:915.
4. Philbrook, D.M. and C.P. Grady. 1985. "Evaluation of Biodegradation Kinetics for Priority Pollutants." *Proc. 40th Industrial Waste Conference.* Purdue University.
5. Eckenfelder, W.W. and Jack L. Musterman. 1995. *Activated Sludge Treatment of Industrial Wastewater.* Technomic. 1995.
6. Hoover, P. 1989. M.S. Dissertation, Vanderbilt University.
7. Mueller, James A., *et al.* 1995. "Fate of Octamethylcyclotetrasiloxane (OMCTS) in the Atmosphere and in Sewage Treatment Plants as an Estimation of Aquatic Exposure," *Environmental Toxicology and Chemistry.* 14(10) 1657–1666.
8. Jenkins, David, Michael G. Richard and Glen T. Daigger. 2003. *Manual on the Causes and Control of Activated Sludge Bulking, Foaming and Other Solids Separation Problems.* 3rd Edition Lewis Publishers.
9. *Wastewater Engineering Treatment and Reuse.* International Edition, Metcalf and Eddy. McGraw Hill. 2004.

CHAPTER 8

Shale Gas Water Management

INTRODUCTION

The development of directional drilling techniques and hydrofracturing has allowed the large scale development of shale gas plays in North America. U.S. gas shales buried thousands of feet below the surface contain over 1,000 trillion cubic feet of natural gas [1], which can now be tapped as a major domestic source of energy. Shale deposits are normally impervious formations, with matrix permeability of 0.01 to 0.00001 [2] milliDarcy (mD) versus 100 mD for petroleum cap rock formations and 5,000 mD for unconsolidated sand) [3]. Hydrofracturing is the process of using water at high pressure to create fractures in the shale, which then serve as conduits to allow gas to diffuse from the rock into the well bore.

Gas-bearing shales are typically between 60 and 90 percent silica minerals, with organic contents of < 1 to > 11% [4]. Many of the shale gas formations in the U.S. (including the Marcellus Shale and Barnett Shale) are the product of the deposition of organic matter from the ocean and organic rich runoff from land that accumulated in deep anoxic troughs during the collision of the African plate overriding the North American plate. The rich organic accumulations were gradually buried under other sediments, and now reside along a wavey line from the Marcellus Shale through Alabama, Arkansas and Louisiana and through Texas. As a result of the marine depositional environment, these formations also contain the salts left by the sea water in the depositional environment. In addition to large quantities of sodium and chloride, gas shales also contain calcium and potassium, and smaller concentrations of barium, magnesium and strontium. Gas shales are generally dry formations, and contain no water in the formation. Formation depth, pressure and temperature information is shown in Table 8.1.

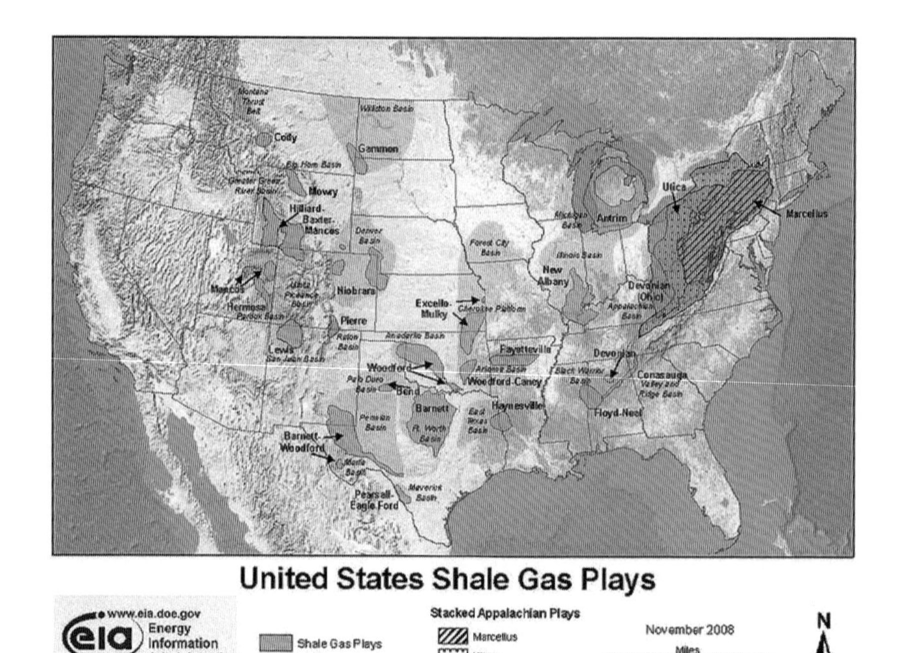

Figure 8.1. United States shale gas plays.

At the point where hydrofracturing activities start, the well has already been installed. The well completion process is completed sequentially in segments from the farthest end of the well lateral. A typical segment length is on the order of 300 to 400 ft for Marcellus Shale wells, and can vary from shale play to shale play. For the initial segment, a perf gun is lowered into the cased wellbore, and set off. The perf gun creates perforations in the steel casing, the concrete surrounding the casing and into the shale. Then water with a prop-

TABLE 8.1. Shale Gas Formation Conditions.

Shale Play	Depth (ft)	Pressure (psi)	Temperature (°F)
Barnett	7,500	4,000	200
Fayetteville	4,500	2,000	130
Eagle Ford	11,500	5,200	335
Haynesville	12,000	8,500	340
Marcellus	7,000	4,000	130
Bakken	10,000	5,600	140

1. Transform Services and Software Web Site – Shale Gas Reservoirs: Similar, Yet So Different, A Comparative Study of North America's Largest Shale Gas Reservoirs.
2. "High Rate Hydraulic Fracturing Additives in Non-Marcellus Unconventional Shales" Rick McCurdy, Chesapeake Energy, presented at EPA Hydraulic Fracturing Workshop #1, February 24-25, 2011.

pant and other chemical additives is pumped at high pressure and flow into the well, building pressure until the formation fractures. When the formation fractures, the pressure drops, and pumping is stopped. A seal is placed in the casing to isolate the segment, and the next segment goes through the process of perforation, hydraulic fracturing and isolation. When the last segment has been fractured, the seals are drilled out, and flowback starts.

In flowback, free water isolated by the seals is released as the seals are drilled out, and comes through the well casing to the surface. This water contains remnants of the chemicals used in hydrofracking, as well as the water that was in the casing and didn't have significant contact with the formation. The flowback is conveyed to a separator tank to allow gas/liquid/heavy solids separation. Initially little gas comes with the water, and over time the gas content increases and the water content decreases. During the flowback period, the concentration of contaminants generally increases with time, while the rate of liquid flow from the well decreases with time. At some point, the flow from the well is almost all gas, and the liquid content drops to a low rate, and the well is placed in operation or shut in for future connection to the gas distribution network.

While reference information is cited for a number of shale plays, most of the treatment and economics discussed in the later portions of this chapter are based on Marcellus Shale experience, and reflect practices and cost structures in that region. The overall thought process is applicable in other plays.

WATER SUPPLY AND PREPARATION

Water used in hydrofracturing is taken from a variety of sources. In the Marcellus Shale, where water is relatively plentiful, water from streams and wells is used to supply hydrofracturing facilities. In the Bakken Shale formation, brackish ground water is treated for use as a water supply, as well as surface supplies hauled significant distances. In the Barnett Shale in Texas, water supply is scarce, and companies have had to resort to more expensive measures, including providing impoundments to store water when it is available, purchasing excess capacity from municipalities, and treatment with portable evaporators to recover water for further hydrofracturing. Producers generally want to have the full water supply that might be needed onsite at the start of hydrofracturing, so that there can be no costly interruptions in completion of the well. Typical water supply requirements for hydrofracturing in several formations are shown in Table 8.2. As shown, a significant amount of water is required to hydrofracture a horizontal well.

A number of additives are required to condition the water for use in hydrofracturing. The largest by mass of these is proppant. Proppant is either specially graded sand or engineering ceramic material, which is carried with the frac fluid into the formation. Its purpose is to prop open the fractures that develop

TABLE 8.2. Shale Gas Wells Typical Water Usage.

Shale Play	Volume of Drilling Water per Well (gallons)	Volume of Fracking Water per Well (gallons)	Total Volume of Water per Well
Barnett	250 to 400,000	2,300,000	2,700,000
Fayetteville	60,000	2,900,000	2,960,000
Haynesville	1,000,000	2.7 to 4,800,000	3,700,000 to 5,800,000
Marcellus	80,000	3.8 to 5,000,000	3.9 to 5,000,000

Basis:
Conversations with Marcellus Shale operators.
Conversations with Haynesville Shale operators.
DOE Modern Shale Gas Development in the U.S. 2008.
James Werline, Devon Energy, EPA HF Technical Workshop, March 2011.
Haynesville Trail Well, December 2011, World Oil Online, Lowry, Yeager *et al.*

in hydrofracturing, to provide an open fairway for the gas to exit the formation into the well bore. Typical sand sizes used for proppant are 20 to 40 mesh. Typical application rates are 0.1 to 1 lb/gallon.

There are two general types of hydrofracturing employed, and they are different due to the method of handling the proppant. In gel fracing, a gelling agent, usually guar gum, is used to distribute the proppant evenly throughout the solution. When a frac stage is completed, a viscosity breaker or gel oxidizer is added to break the gel so that it doesn't restrict the flow of gas. In slickwater fracing, friction reducers are used to allow for pumping at very high velocities, which are sufficient to carry the sand into the fracture zone uniformly. Gel fracs typically result in larger primary fractures, whereas slickwater fracs producer a greater distribution of finer fractures. Hybrid mixtures of gel additives with friction reduction have been used as well.

Other additives to the hydrofracturing water supply at injection include biocides, scale inhibiters, pH adjustment, corrosion inhibiters, and specific additives for iron control, emulsion prevention, and in the case of gel fracs, oxidizing breakers and enzyme breakers. In a typical frack fluid, water is over 90% of the mass introduced into the well, with proppant approximately 9% typically. The remainder, less than 1%, includes all the chemicals added for the various functions described.

The friction reducers most commonly used today are polyacrylamide polymers. These polymers are also used in wastewater treatment for flocculation. In hydrofracturing they are used at much higher concentrations. They are typically used in concentrations of a few ppm for wastewater treatment, and 500 to 1000 mg/L in hydrofracturing. These polymers are capable of withstanding the temperatures in the formation, and are generally compatible with the other chemicals used in hydrofracturing. The goal of friction reduction is to reduce the friction pressure loss by about 70% over that associated with untreated water. This is critical because of the pressure in the formation, discussed later, and the need in slick water fracking to maintain velocity to maintain the sand in suspension. This is balanced by the need to use production casing wall thicknesses that are (1) commercially available, and (2) compatible with the forces exerted in the horizontal drilling process. Typical burst pressure for commercial production casing is on the order of 8800 to 13200 psi.

Polyacrylamides are limited by salt content, iron, calcium and magnesium concentration. Typically cited criteria for water supply compatibility with polyacrylamide are shown in Table 8.3. Salt content (TDS) can cause breakdown of polymers used in friction reduction. Recent developments in this area have yielded polymers that are effective in 10% to 12% sodium chloride brine [5], which allows for the recycle of frac flowback water with pretreatment.

Friction reduction is evaluated using a friction loop. This device consists of a small tank, pump (frequently progressive cavity type) and small diameter tubing which mimics the velocity and Reynolds Number in the well production casing, and produces significant friction loss as a result. Initially water or a frack fluid is pumped around the loop without the addition of a friction reducer. Pressure is recorded at several points along the loop. Following that, the dosage of one or more friction reducers is varied, and pressures are again monitored. This approach is used to determine the dosage to be used in the field. Additionally, this type of apparatus allows for evaluation of possible impacts on friction reducers from biocides and other frack water additives.

While there are some polymers advertised as being non-adhesive to the shale, it may be necessary to break the polymer chemically to prevent restriction of gas movement. Oxidizing agents (frequently persulfates) are used to destroy polymer adsorbed to the shale, which would otherwise tend to reduce gas production. In gel fracs, ammonium persulfate is frequently used to degrade the gel for the same purpose.

Another consideration involves biological activity potentially fouling the well. Principal bacteria of concern include sulfate reducing bacteria, anaerobic acid formers, and aerobic slime-forming bacteria (more of a concern above ground and in the well bore than in the shale formation)[6]. Sources of oxygen for anoxic bacteria include nitrates and sulfates. While sulfates are generally

TABLE 8.3. Polyacrylamide Sensitivity Limits.

Contaminant	Units	Limitation
Ferrous Iron	mg/L	< 5
Manganese	mg/L	Varies
Calcium	mg/L	< 1,000
Magnesium	mg/L	< 1,000
TDS	mg/L	5% to 12%*

*Function of friction reducer used.

low in the formation, there may be sulfates present in the fresh water source used for fracking. Nitrates may be present in explosive residue. Organic substrate sources include friction reducers and other additives to the frack water from the frack water recycled for reuse. The primary concern associated with biological activity is the formation of slimes and souring of the well from biological activity in the formation.

Key criteria for evaluating disinfecting agents are efficacy of bacteria destruction, interaction with friction reducers and other chemicals in the frack water and environmental persistence. It is generally viewed as important to be able to measure a residual in the field, particularly in the flowback water. While oxidizing biocides are generally effective, and easily measurable, interaction with other components in the water, including the aforementioned chemical additives as well as ammonia, may make them unattractive for downhole use. Disinfection of raw water supplies for fracking has been accomplished with chlorine dioxide and ozonation systems. Ozonation mobile systems have been used to degrade organic matter in flowback water as well as providing disinfection for flowback water [7].

A number of non-oxidizing bactericides have been used in Shale Gas plays as a component of the hydrofracturing fluid. One commonly used biocide is glutaraldehyde. Glutaraldehyde has the chemical formula $CH_2(CH_2CHO)_2$. Other uses include disinfection of medical and dental equipment, and biological control in a variety of industrial water systems. Glutaraldehyde kills bacterial cells through cross-linking proteins. It does not interact with common friction reducers, antiscalants or other hydrofracturing water additives. Typical concentrations of glutaraldehyde in frack water are on the order of 500 mg/L. Colorimetric test kits are available to test residual in the field. Other amine and carbamate non-oxidizing biocides are also used as downhole biocides.

Caution is required in selection of biocides due to possible interaction with the friction reducer. In the case of using anionic polyacrylamide as a friction reducer, use of cationic quartenary amines either separately or in combination with glutaraldehyde has been shown [8] to cause a reduction in friction reducer efficacy. Friction loop testing provides insight into the reactions between components such as biocides and friction reducers.

Scaling in the formation is also a major concern of producers. As there is no control over the dissolution of salts in the formation that may lead to scaling, the key approach is to minimize the constituents of water used for fracing to control scaling. Thus, water supplies that are high in sulfate typically are not attractive for fracking, because of the potential of the sulfate in the mine drainage to precipitate barium sulfate, or in extreme cases, strontium and calcium sulfate. Since calcium, magnesium and strontium are typically dissolved from the shale, scale inhibiters are normally added to minimize the formation of carbonate scales in the event that saturation conditions are reached for scaling compounds.

Other additives in fracking include emulsion preventers, iron chelators, and various carrier components for the chemicals noted. It is typical to see ethanol, methanol and ethylene glycol as components of chemical mixtures used in fracking, in low quantities, as carriers for other components.

FLOWBACK WATER QUALITY AND QUANTITY

Once the well hydrofracturing has been completed, the completion company begins to drill out the seals, which allows the flow of water and gas to the surface from the well. Typical depths and formation pressures in gas shale formations are shown in Table 8.1. As shown, depths can be over two miles to some shale plays. Formation pressure is the pressure in the well bore after fracturing, and as shown is significant. In comparison to the Barnett Shale formation pressure shown, a column of water 7550 ft high would exert a pressure of about 3250 psi. Thus there is more than adequate formation pressure to move the water in the production casing and near the well bore out the top of the well.

When flowback starts, the rate of water and gas release from the well is restricted through valves at the surface. Initially, much more gas than water is released from the well, and over time, the flow of water decreases and the flow of gas increases. At some point in the process, the completion company determines that the flow from the well is mostly gas, and can at that point put the well on line (i.e. connect to gas collection piping) or shut the well off if demand or pricing for gas is low, or if collection piping has not reached the well site yet. The period in which mostly water is conveyed from the well is termed the flowback period, and the water generated during this time is flowback water. After the flowback period is completed, water from the well is termed produced water. The term originates from the oilfield situation in which water and oil are brought from the well and separated, with the water portion being referred to as produced water. In this case, the produced water is not produced by the formation, but is water from hydrofracking that is released slowly over time from deep inside the fractures of the formation. Figure 8.2 shows typical data from Marcellus Shale wells for flowback water flow rates. As shown, the period where flow rates of water from the well are high is typically only a few days, after which the flow rate of water gradually decreases with time. For the wells shown, by the time two weeks have elapsed, the flow of water from the well is less than 2000 gallons/day in all cases.

An important point to keep in mind related to the return flow of water from hydrofracturing is that not all the water pumped into the formation in hydrofracturing returns to the surface. In the Marcellus Shale play, most producers indicate the long term return of fluids from the well as flowback water and in the first few months of produced water totals under 30% of the applied water.

From personal observation and experience, the range of return flows can be as low as 10% to above 50%, but both are rare compared to a more normal 20% to 25% return of flowback water. In the Barnett Shale, DOE reports suggest a return rate of about 1/3 of the applied water in hydrofracturing.

While the volume of flowback water from the well decreases with time, the concentration of contaminants in the flowback water increases with time. When the flowback period starts, and water first flows from the well, the water coming out of the well has had very little contact with the shale formation. As hydrofracturing water continues to flow, water with more contact with the formation starts to come from the well, and the concentrations of contaminants increase. In wet formations, water from the formation will also increase with time, and will contain contaminants as well. Finally, when the well is in production, the produced water will have had long term contact with the formation, and will be essentially in equilibrium with the formation. For the Barnett and Marcellus Shales, this is easily seen in the plot of dissolved solids versus time for flowback shown in Figure 8.3. This curve is a composite of data from several Marcellus Shale wells, and reflects the increasing TDS concentration from low values at the start of flowback increasing to that of seawater (3%) in under 30 hours, to over 8% TDS in 6 days. Observed data from Barnett Shale wells is similar in the overall trend. Depending on the formation characteristics and flowback flow rate and the number of frack stages, higher or lower rates of increase can be observed.

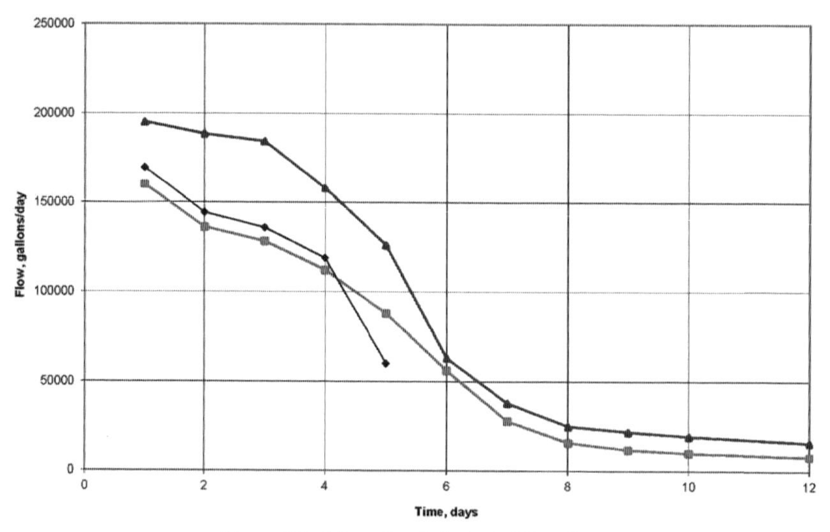

Figure 8.2. Marcellus shale typical flow data.

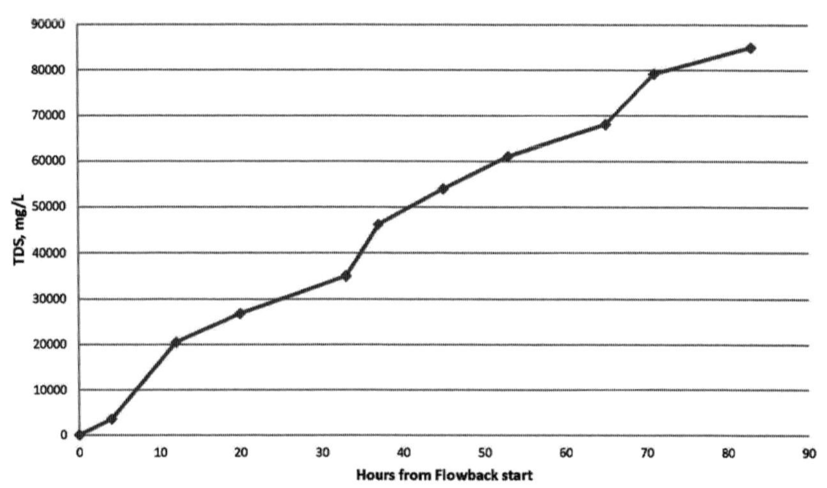

Figure 8.3. Marcellus frac flowback TDS composite plot.

FLOWBACK WATER CHARACTERISICS

Understanding the chemistry of hydrofracturing flowback and produced water is critical to understanding the potential types of treatment or disposition for these waters. The chemistry of frac water results from both the chemicals added to the hydrofracturing fluid and the dissolution of constituents of the shale.

1. Salt

Unconventional Gas Shales frequently have substantial amounts of readily solublized salts, included sodium chloride and potassium chloride. These originate from the marine environment in which the material was originally deposited. The TDS levels in Marcellus Shale wells typically are relatively low at the start of flowback, but increase with time, as reflected in Figure 8.3, above. At the end of the flowback period, sodium chloride levels can be close to 200,000 mg/L. Produced water samples from Marcellus Shale wells have been observed to range between 210,000 and 260,000 mg/L Similar TDS levels can be observed in the Bakken Shale [9], with faster increase in salt/TDS levels in the Fayetteville and Haynesville Shales [10]. In comparison, data from the Barnett [11] and Fayetteville shale indicate lower rates of salt concentration increase. Extensive review of frack water data indicates that about 90% of the dissolved solids in hydrofracturing water are typically sodium, potassium and chloride. Potassium levels vary but are generally small compared to sodium concentration.

2. Scalants

Scalants are those cations that can react with commonly present anions in water to form scale on surfaces in contact with the liquid. The critical parameters in hydrofracturing are calcium, magnesium, barium and strontium. All are present to some extent in shale gas plays, and all are solublized in hydrofracturing, although there is considerable variability in concentration in flowback and produced water, even within the same play.

In the Marcellus Shale, Barium in flowback water has been noted as varying from concentrations in the low hundreds of ppm in southwestern PA to over 5,000 mg/L in wells in northeastern PA. A well in Bradford County contained over 13000 mg/L Ba in late flowback water [12]. Characteristics of mid-flowback or composite flowback water from wells in these areas is shown in Table 8.4, below. Observation of the data indicates that barium can range from low values to over 5000 mg/L. Barium also is a good example in this case of spatial variation within the same play, as Barium concentrations tend to be lower in Southwestern Pa., and higher in Northeastern Pa. However, even within the area of Southwestern Pa., significant variation can occur within the space of a few miles.

Calcium concentrations in composite or mixed flowback samples vary in the table from under 1000 to over 7500 mg/L. Calcium and the other scalants tend to parallel the TDS curve shown in Figure 8.3 in terms of increasing concentration with time. Water at the end of the flowback period has been observed in excess of 30,000 mg/L Ca. Produced waters have been observed in the Marcellus Shale at over 50,000 mg/L Ca. Less extreme variations have been observed in magnesium and strontium concentration.

Other shale plays experience the presence of these same constituents, in varied amounts. Where TDS levels are lower, generally the concentrations of barium and calcium are lower as well.

TABLE 8.4. Flowback Water Scalant Composition Marcellus Shale.

Parameter	Davis Well, Marshall Co WV, Day 5 Flowback, May 2009	Southwest, PA, February 2009, Day 3+ Flowback	Williamsport, PA, Composite Sample	Susquehanna Co, Composite
TDS, mg/L	100,000	105,000	110,000	40,000
Ca, mg/L	7,630	7,500	6,896	736
Mg, mg/L	829	640	725	127
Sr, mg/l		1,700		228
Ba, mg/L	136	170	5,145	596
Fe, mg/L	38	24	39	8

3. Suspended Solids

Flowback water generally carries some sand as well as finer suspended solids resulting from the perforation of the well casing and the fracture of shale. The water from the well head in flowback is conveyed to a sand/oil/water separator, to remove the heavier solids and any oil present. The liquid is then conveyed to storage, most frequently in frac tanks or lined impoundments. This presents another opportunity for settleable solids to fall out of suspension, and generally, there is some residual solid material in the bottom of the frac tank after use in this service. The suspended solids remaining after all this it typically in the 200 to 500 mg/L range for Marcellus Shale and similar plays. Samples within the first few hours of flowback may be slightly higher. In the longer term, samples of produced water in the Marcellus Shale are typically 300 to 400 mg/L TSS, although occasionally a sample over 1000 mg/L TSS may occur.

4. Organics

Shale gas formations exist in three forms for the subject of organics in flowback water, and there are three types of wells as a result. Natural gas wells produce methane, but not appreciable amounts of ethane, butane, pentane or larger organic molecules. These are generally referred to as dry gas wells. The Marcellus Shale is for the most part a dry formation, with the exception of its western-most end as shown in the attached map, Figure 8.4. The western end of the Marcellus Shale has condensate wells, in that it pro-

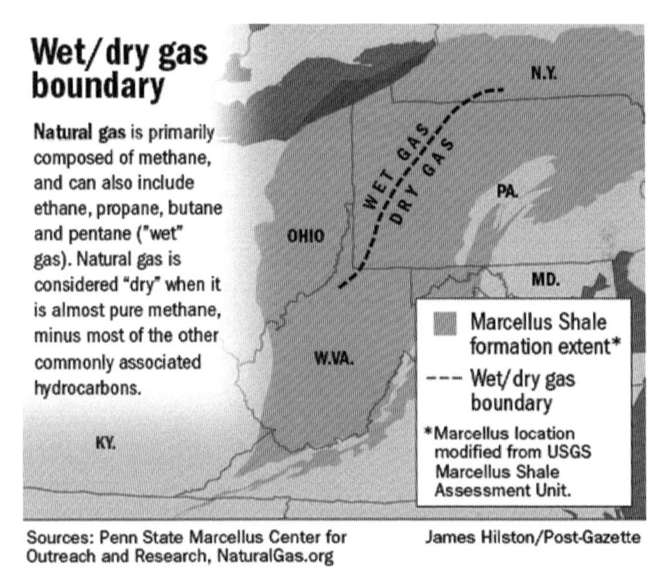

Figure 8.4. Wet/dry gas boundary.

duces gas with liquids. Liquids are defined as ethane, propane, butane and larger organics. Liquids are liquid under the pressures they experience in the formation, but are gaseous at ambient pressures and temperatures. Liquids or condensate are generally removed or recovered from condensate wells, because (1) they add value, and (2) the condensate portion has a higher BTU content than methane. The natural gas industry requires that input to the natural gas distribution system be at a consistent range of BTU values, so that homeowners don't experience damage to natural gas burning equipment as a result of receiving spikes of higher BTU content material. The U.S. standard for natural gas is 1035 BTU/cu ft gas at 1 atmosphere and 60°F. A third type of well is referred to as an oil well, which produces both oil and natural gas. The Bakken Shale is an example of a field of oil wells, as oil is the principal product of the Bakken Shale, although substantial amounts of gas are also present.

In wastewater treatment, organics are frequently measured using tests that quantify the total organic content based on oxygen demand(biochemical oxygen demand, BOD_5, or chemical oxygen demand, COD), or oxidation of organic material to carbon dioxide with measurement of the carbon dioxide(total organic carbon or TOC). It is difficult to rely on these tests for meaningful results. BOD_5 measures the consumption of oxygen by bacteria. To avoid detrimental effects from the salt content and potentially toxic materials, significant dilution is required, which would yield poor accuracy. The chemical oxygen demand test measures the consumption of a strong oxidizing agent. One of the potential interferences for this test is chloride. Chloride can be oxidized to chlorine in the test procedure, and while small amounts are not generally an issue, at the concentrations normally found in this type of sample, the interference provides a high reading. Organic carbon analysis, which involves oxidation of organics to carbon dioxide followed by measurement of carbon dioxide, is likely the best method of the three for detecting gross organic content. Oil and Grease analysis is also used, which involves hexane extraction of organics. However, other organics can be extracted as well. Experience in the dry portions of the Marcellus Shale indicates that Oil and Grease concentrations in the raw flowback water are relatively low, in the range of 0 to 20 mg/L, based on limited testing.

Consideration should be given in dry formations to the types of organics applied in the frac water. These can include concentrations of up to 1000 mg/L if friction reducer (typically polyacrylamide), up to 500 mg/L of disinfectant (frequently glutaraldehyde), dispersants, surfactants and scale inhibiters, and carrier organics including ethanol, methanol and glygols in the range of 100s of mg/L. While some of this may be lost in the formation or adsorbed to the surface of the shale, much would be expected to return in the frac water. From condensate wells, there may be some contribution from higher organics, although most would be expected at wellhead pressure to be in the gaseous phase. In the case of oil wells, there can be appreciable amounts of organics from the oil as well as the normal constituents in the water.

5. Metals

In the Marcellus Shale, substantial data has been collected by a variety of entities related to the characteristics of flowback water. One such effort is outlined in the work done by Tom Hayes of GTI, referenced previously. The general consensus from these efforts is that metals represent a very small constituent in flowback and produced water. The data quoted in Mr. Hayes data indicated average values for all metals was well below 1 mg/L. Occasional peaks of a few ppm of zinc and copper were noted. Two values not mentioned in this work are iron and aluminum. Aluminum is frequently present at a concentration of a few mg/L. There is also generally a low concentration of fluoride present, which may allow for the aluminum to exist in soluble form at pH levels under 6.5. Iron is also generally present, and has been observed by many researchers as increasing with time during flowback. Typical values of 20 to 40 mg/L are indicated early in the flowback of Marcellus wells, with increasing iron to over 80 mg/L toward the end of flowback. Observation of a number of analyses for produced water from the Marcellus Shale generally indicates that while the peak total iron may be higher, it is rare to see soluble iron above 200 mg/L.

6. Radionuclides

Naturally occurring radioactive material is normally present in shale. Some have attributed this to the bioaccumulation of radioactive compounds by the once living animal and plant life whose decay produced natural gas. Gamma ray logging is a frequent component of exploration wells, and correlates with gas content in shale. The EPA [13] indicates that the three components of concern related to natural gas wells are Radium 226, Radium 228 and radon. When flowback water contains these compounds, they report to the flowback water handling facilities. Radium is a homologue of barium, as they are both alkaline earth metals. Chemically they are similar in terms of solubility. Historically, there have been barium scales observed to accumulate in well casing. There is a tendency for radium to precipitate with barium in these scales, and they typically exhibit significant radioactivity [14].

Radioactivity is controlled at Pennsylvania landfills through monitoring of the trucks entering the landfill. There are two radioactivity levels evaluated for each vehicle entering a landfill. A fixed radiation sensor is used to monitor radiation on incoming vehicles. The criteria followed is that if the reading from the sensor is under 10μR/hr above the background level, the load is accepted. If the load is greater than 10 μR/hr above background, the load is pulled to the side, and inspected more thoroughly. The criteria then applied is that the vehicle must be further inspected and evaluated. Criteria exist for permissible radiation dosages in the vehicle cab and the sides of the vehicle. If the vehicle is under 2 mR/hr in the cab and under 50 mR/hr on all other surfaces, it can be accepted at the landfill's discretion (landfills have limited ability to do this). If the vehicle is over either of the two criteria indicated, the NRC is called in, and the material requires handling as a nuclear waste. Conversation with Pa. DEP personnel in the past indicated that this was an extremely rare event.

Characterization has been done in the past at a number of Marcellus Shale wells for radium content. However, use of this information requires an understanding of further treatment of the frac water, and translation from picocuries of radium to a rate of emissions. Given that, it does not seem that this facet of flowback water currently is an issue.

7. Other Constituents

A variety of other constituents show up in hydrofracturing flowback and produced water, and are worth consideration.

- *Ammonia:* Ammonia has been observed to be present in flowback water when evaluated. The range observed in the Hayes reference was from the single digits to over 400 mg/L.
- *BTEX (Benzene, Toluene, Ethylbenzene and Xylene)*: Initially checked in early testing and found to be very low in Marcellus Shale flowback. More likely in condensate well returns and oil well return flows.
- *Silica:* Observed values of 20 to 100 mg/L as total silica in Marcellus Shale samples
- *Other parameters that can be present in lesser amounts:* Boron, Bromide. Fluoride, Lithium, Manganese

HYDROFRACTURING WASTEWATER DISPOSITION

How flowback water and produced water is handled depends on a variety of factors that have cost implications. The problem is complex, and the remainder of this chapter will focus on some of the options, and why they are attractive. From the preceding sections, we understand that a great deal of water is used in hydrofracturing, but only a fraction returns to the surface. The water that does return to the surface contains high concentrations of sodium chloride (and some potassium and bromide), potential scale formers in barium, calcium, strontium and magnesium, iron and possibly some manganese, some relatively small amount of radium, and some other components in lesser amounts.

Initially, in the Marcellus Shale region, the historical practice was followed of trucking flowback water to a nearby publicly owned treatment works for treatment and discharge. As seen in the description of flowback characteristics above, the salinity of flowback water is high. POTWs don't provide effective removal for the soluble salts, or for dissolved strontium, barium, radium, etc. In late 2008, TDS levels above the recommended water quality criteria for drinking water were initially observed at several locations in the Monongahela River basin, which is the drinking water source for a number of communities in northern West Virginia and western Pennsylvania. Following investigations by numerous parties, it was determined that while not the entire cause of the problem, hydrofracturing water was a component of the problem, and should not be discharged from municipal treatment facilities. Producers were asked to voluntarily discontinue sending frac water to pretreatment or municipal facilities in early 2011, and they have complied.

In the western U.S., the salinity of drinking water supplies is a more critical issue than in the east, and disposition of flowback waters at POTW facilities isn't permitted.

In addition to the considerations of pass-through pollutants and the potential introduction of radium into surface streams, many POTW operators became concerned with the impact of high chloride concentrations they experienced when accepting flowback water on their concrete, steel and cast iron wetted components in the treatment facility.

The simplest method available for disposal of frac flowback water is deep well disposal. Expensive treatment is avoided, and pretreatment costs are minimal. This option is only available where the geology is compatible, and the regulatory agencies permit it. In Texas, this option has been used extensively. In Pennsylvania, this option is rarely used. The availability of this option varies from state to state. Two additional factors need to be considered. One is the cost to transport the wastewater to a disposal site. The other is the need for fresh water for fracing in dry climates. Some combination of these factors makes the use of evaporators attractive to produce a concentrated brine for disposal, and fresh water for more hydrofracturing. Costs will be discussed in a section on current economics following this section.

In areas where deep well disposal is not an attractive option, the approach was developed that flowback water from one hydrofracturing event would be used in the next hydrofracturing event. This occurred in the Marcellus Shale between 2008 and 2011. In 2008/9 most of the hydrofracturing water was conveyed to municipal treatment plants and brine treatment plants, both of which removed suspended solids and iron and discharged to a surface stream. In 2010, much of the water was recycled for reuse. Due to the impact of dissolved salts from hydrofracturing on the drinking water supply of large portions of Pennsylvania and West Virginia, the State of Pennsylvania requested that all producers cease taking water to municipal treatment and brine treatment facilities in March 2011. From that point forward, nearly all flowback water generated was either used in recycling or trucked to other states for deep well disposal. In addition to requesting that producers not take water to the indicated facilities, the state of Pennsylvania also enacted revisions to 25 Pa Code Ch 95 in August 2011 restricting the TDS from wastewater generated in the oil and gas industry to 500 mg/L, with associated restrictions on chloride (250 mg/L), barium(10 mg/L), and strontium (10 mg/L).

The limitations on flowback reuse in the Marcellus play prior to 2009 were primarily focused on the following limitations:

1. Friction reducer was sensitive to TDS level, and at that time could tolerate only a couple percent TDS.
2. Friction reducer was sensitive to calcium and magnesium.
3. Friction reducer could tolerate only a few ppm of iron.
4. Scaling conditions could occur from high barium, dissolved iron or calcium concentrations.
5. Suspended solids could cause issues with pore blockage in the formation.

Additionally, there were environmental concerns related to storage and handling of flowback water. From the point at which recycling became an industry focus, consideration was given to the level of pretreatment required to alleviate the concerns shown above. The most critical limitation was that of the friction reducer for dissolved solids. As noted previously, friction reduction is absolutely critical for slickwater and hybrid fracs, as the velocity in the pipe has to be maintained to keep the sand suspended. Developments by chemical suppliers yielded a modification of the polymer formulation that provided enhanced compatibility with dissolved salts. In 2009, friction reducers were tested that could tolerate in excess of 10% dissolved solids [15]. The observed characteristic of Marcellus Shale wells as noted above is that about 20% to 30% of the water applied in a frack returns as flowback. This indicates that even if the flowback and produced water from a given well together average close to 20% TDS, by the time they are diluted with fresh water 4/1, the resulting mixture will be only about 5% TDS, which is certainly tolerable from a friction reduction standpoint. The other issues could be addressed through conventional wastewater treatment processes.

Through experimentation, Marcellus Shale producers determined the treatment needs for water to be recycled. There has been no industry standard for flowback water treatment to date. Some producers have used flowback water without pretreatment. Most require a minimum of iron and suspended solids removal prior to reuse. Many include barium removal as a requirement. Some are reported to use oxidation for organics breakdown as well as iron oxidation. Some have discussed reducing the calcium concentration with softening in areas where very large calcium concentrations occur in flowback water.

Iron oxidation is frequently done using aeration. The stoichiometric relationship [16] is:

$$4\ Fe(II) + O_2 + 8\ OH^- + 2H_2O = 4\ Fe(OH)_3$$

As shown, hydroxide alkalinity is consumed, which generally requires the addition of an alkali to maintain pH levels. The reaction occurs readily above a pH of about 3.5, resulting in the precipitation of the ferric hydroxide floc. The source of oxygen is generally from aeration of the water being treated.

The precipitation of barium is also required in some cases. Barium sulfate has a very low solubility. Barium sulfate is also an attractive precipitant because it is stable and can pass the TCLP test for barium. Barium is a concern for producers due to the low solubility of barium sulfate ($K_{sp} = 1 \times 10^{-10}$). The solubility of barium sulfate in the presence of equal molar amounts is only about 2.4 mg/L $BaSO_4$. While there may be little sulfate in the formation, there is barium present. As barium dissolves from the shale, it contacts frack water which is partially fresh water, and would be expected to contain some sulfate. In areas in which coal mining was conducted in the past (i.e. most of the Marcellus Shale regions in Pennsylvania and West Virginia), sulfate concentrations in streams contaminated with AMD can be over 100 mg/L. Given an excess of barium in the formation and 100 mg/L sulfate in the frack water, the resulting precipitation reaction could cause almost 250 mg/L barium sulfate precipitation, which would have a significant impact in terms of scaling the shale, and reducing gas flow. Barium sulfate has historically been observed as scale forming inside shale gas well casings.

The relationship between cations and sulfate in terms of precipitation sequence is illustrated in Figure 8.5, following this page. This figure represents the output of a Visual Minteq model of the reaction chemistry for sulfate precipitation given normal frac water characteristics. The model is outside the range of accuracy noted in the Visual Minteq model, but has been compared with bench scale test results and appears to be reasonably accurate. The results demonstrate that as a sulfate source is added, barium is initially precipitated. The sulfate concentration in solution stays low until nearly all of the barium is precipitated, at which point the sulfate concentration increases to the point where strontium starts to precipitate. In turn the precipitation of strontium sulfate continues until nearly all of the strontium is precipitated. Note that the secondary y axis does not extend to zero, but only covers the beginning of calcium sulfate precipitation.

The precipitation of barium sulfate is a function of barium and sulfate concentration, not pH. The low cost choices of reagent are sulfuric acid and barium sulfate. Both have been used in the past to precipitate barium sulfate, and both are effective. The choice between them should be driven by chemical cost (including an alkali to neutralize the sulfuric acid), and safety/risk concerns. Barium sulfate is notorious for scaling. In an improperly designed reaction

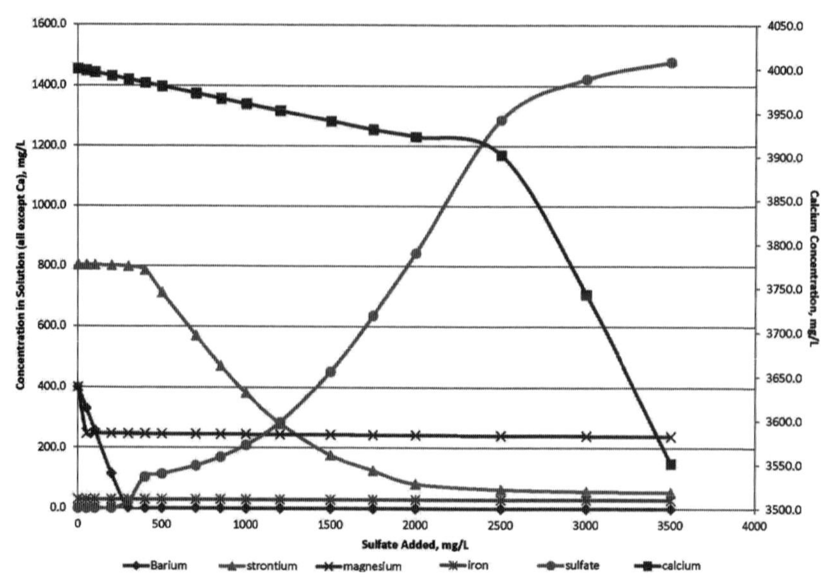

Figure 8.5. Frac water soluble ions vs. sulfate addition.

tank, the solids can scale on the walls gradually reducing the effective tank diameter. Barium sulfate scale is fairly hard, like calcium sulfate scale.

The difficulty in control of barium sulfate precipitation lies in the ability to detect the barium or sulfate concentration. Barium is only measured individually using fairly expensive and sophisticated equipment. While makers of portable test equipment have developed field tests for the combination of barium and strontium, there is not at this time a field test that can be used to quickly and effectively determine the barium concentration in the presence of strontium. There is also no ion sensing electrodes that will function in the presence of strontium, and no convenient method to remove strontium prior to testing with either a field kit or a ion specific electrode.

Once precipitated, the solids produced will settle effectively by gravity. The solids concentration will be excessive for a plate type settler, and reaction clarifiers or conventional wastewater clarifiers are preferred. Generally, the solids produced is sufficient to push the clarification regime to hindered settling. For a given producer with a set hydrofracturing chemistry and technique, it is best to do bench and if possible pilot testing to identify design criteria for a full scale installation. Barium sulfate solids are effectively dewatered in filter presses.

A number of chemical oxidizers have been used in the field for oxidation of organics, including chlorine dioxide, peroxide and persulfates. The assessment of need for a chemical oxidizer can depend on the amount of friction reducer returning in the flowback water. If enough solids are precipitated, the friction reducer polymer may adhere to the solids. If there is too much friction reducer remaining, the resulting charge balance may stabilize small particles in suspension, making gravity settling or filtration difficult. Testing is required to evaluate the need for oxidant addition.

In some cases it may be desirable to remove some calcium and magnesium. The chemistry of these reactions was also evaluated using Visual Minteq, as above. The resulting model is shown as Figure 8.6. As shown, calcium tends to hog the reagent carbonate to a point where the majority of strontium has been precipitated, at which point calcium and strontium compete for carbonate reactant. This implies that to effectively remove one to a low concentration would require removing both. On the other hand, if the issue is removal of a portion of the calcium to achieve compatibility with a friction reducer, then partial removal may be accomplished with calcium carbonate precipitation only. pH adjustment to liberate the carbonate in solution as CO_2 following removal of the solids should be considered to minimize the scaling tendency of the recycle water.

In comparison to the eastern shale plays where water is relatively plentiful and disposal is difficult, the general problem in western shale plays is that water is scarce and expensive, but flowback water is relatively easily disposed of via deep well. In the western shale plays, the emphasis is in reducing costs through recovery of water for reuse. A number of suppliers have provided systems to evaporate a portion of the flowback water, recovering clean condensate for reuse in subsequent fracs. In some portions of the western U.S. conditions are suitable to allow for gravity evaporation of the concentrate in evaporation

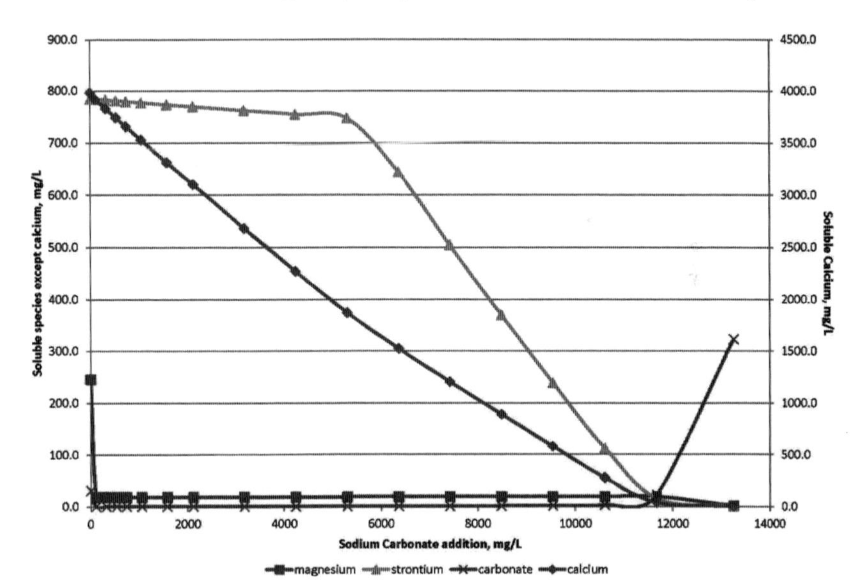

Figure 8.6. Frac water softening soluble species.

ponds, without recovery.

Consideration has been given to the use of reverse osmosis in the recovery of clean water for reuse, particularly from the initial return flowback water. However, to maintain membranes in operation consistently would require pretreatment for iron and scalant removal, or other relatively sophisticated pretreatment. The duration of time in which flowback water could be directed to an RO system is limited, as the TDS level tends to increase relatively rapidly in many plays. The current upper limit for feed to a sea water RO is about 4% salt.

At the other end of the spectrum, evaporation and crystallization processes can be used to produce water for reuse, with no discharge other than salt. There are several options available that involve different end results, and each has different pretreatment requirements. On the most simple level, the use of an evaporator for concentration of brine followed by shipment offsite for disposal is an alternative. If deep well disposal capacity is available within a reasonable distance, this approach could minimize trucking costs for disposal. The ultimate brine TDS from the evaporator discharge would likely be in the vicinity of 25 to 26%. Thus, if the input brine averages 6%, the volume reduction would be to a little less than one quarter the original volume. The pretreatment that would be required for this option could include barium and suspended solids removal. The condensate could be reused for further hydrofracking. Additional treatment of the condensate may be required for organics that carry over from the evaporator.

A second alternative for evaporation is the use of evaporation/crystallization to produce a sodium chloride cake for either disposal or reuse as road salt. In its simplest form, this would include pretreatment as above, operation of the evaporator as above, and conveyance of the concentrated brine to a crystallizer. In the crystallizer, sodium chloride salt would be crystallized. Condensate would be recovered from the overhead stream. A purge stream would be required to remove calcium, magnesium, strontium, metals and other salts to produce a reasonably pure sodium chloride. The purge stream would require disposal, which could be by deep well. The volume of the purge stream would be much lower than the volume of concentrate produced by the evaporator alone. Alternatively, it may be possible to crystallize a second component of value from the purge stream, depending on the makeup of the purge stream. However, this would require a larger capital investment.

A third option is to soften the feed to the evaporator crystallizer, and remove the calcium, magnesium and strontium as precipitates. This approach produces a large amount of solids requiring disposal, but disposal can be in a conventional landfill.

For systems producing sodium chloride salt, the ASTM specification for road salt requires 95% purity, less than 1% moisture, and a particle size range of 1/2″ × No 30 sieve. This generally requires drying of the crystals removed from the crystallizer, followed by processing to achieve particle size distribution specified. Some public agencies require salt supplied for road salt to be produced from salt mining operations or recovery from seawater, but there have been instances in which salt from wastewater treatment operations has been used as road salt.

ECONOMICS

The low cost options at this time for disposition of flowback water are generally deep well disposal, partial recovery with reuse, or pretreatment and reuse. Deep well disposal typically has cost components of transportation to a deep well and the fee for the well operator to accept the water. Typically cited values in the Marcellus Shale play for these costs are $2/bbl (1 bbl = 42 gallons) for disposal, and $100/hour for a 150 bbl truck [17] Truck costs are both ways. If a disposal well is 2 hours away, the bill would be for 4 hours use plus time waiting to unload and unloading. In comparison, the costs associated with pretreatment and reuse are typically cited as being in the range of $0.04 to $0.10/gallon ($1.68 to $4.20 per bbl) for iron and barium removal. This is comparable to the costs charged by brine processors in Pennsylvania, which were in the vicinity of $0.05 to $0.07/gallon ($2.10 to $2.94/bbl). Neither of these figures includes transportation, and the same rates as above should apply.

In comparison, the state of Pennsylvania estimated the cost of treatment of high TDS water (via evaporation/crystallization) at $0.12 to $0.25/gallon (or $5 to $10.50/bbl)[18]. This is less than private sources have projected.

REFERENCES

1. *Modern Shale Gas Development in the United States, A Primer.* U.S. Department of Energy, Office of Fossil Energy, NETL. Executive Summary, ES-1. April 2009.
2. Ibid., page 30.
3. Freeze, R. and J. Cherry. 1979. *Groundwater.* Prentice Hall.
4. Sageman, B.B. and T.W. Kyons. 2004. "Geochemistry of Fine Grained Sediments and Sedimentary Rocks." *Treatise on Geochemistry,* Vol. 7. Elsevier Publishing, New York. P. 115–159.
5. Houston *et al.* 2009. SPE 125987. "Fracture Stimulation in the Marcellus Shale—Lessons Learned in Fluid Selection and Execution." Superior Well Services. Presented September 23–25 at the 2009 SPE Eastern Regional Meeting, Charleston, WV.
6. Tischler *et al.* May 2010. SPE 123450. "Controlling Bacteria in Recycled Production Water for Completion and Workover Operations. *SPE Production and Operations Journal.*
7. See internet publications from Kerfoot Technologies, Fountain Quail Water Management LLC, Ecosphere and others.
8. Rimassa *et al.* 2009. SPE 119569. "Are You Buying Too Much Friction Reducer Because of Your Biocide.", 2009 SPE Hydraulic Fracturing Technology Conference.
9. Harju, John. September 2009. "Bakken Water Opportunities Assessment." Northern Great

Plains Water Consortion at the North Dakota Petroleum Council Annual Meeting.

10. Blauch, Matt. March 2011. "Shale Frac Sequential Flowback Analyses and Reuse Implications." Superior Well Services. Presented at EPA Technical Workshop.
11. Hayes, Tom, GTI. 2011. "Characterization of Marcellus and Barnett Shale Flowback Waters and Technology Development for Water Reuse." Presented at Hydraulic Fracturing Technical Workshop #4, Arlington Va., March 30, 2011.
12. Silva, James. 2011. "NORM Removal from Hydrofracturing Water." IWC 11-07, International Water Conference.
13. U.S. EPA—website Radioactive Wastes from Oil and Gas Drilling.
14. Silva, James. 2010. "NORM Removal from Frac Water in a Central Facility." IWC10-65, International Water Conference.
15. Houston *et al.* 2009. SPE 125987. "Fracture Stimulation in the Marcellus Shale-Lessons Learned in Fluid Selection and Execution." Superior Well Services. Presented September 23-25 at the 2009 SPE Eastern Regional Meeting, Charleston WV.
16. Stumm, Werner and G. Fred Lee. February 1961."Oxygenation of Ferrous Iron." Harvard University, *Industrial and Engineering Chemistry.* Vol 53, p 143.
17. Rassenfoss, Stephen, JPT. July 2011. "From Flowback to Fracturing: Water Recycling Grows in the Marcellus Shale." SPE online print archive.
18. "Marcellus Shale Wastewater Issues in Pennsylvania-Current and Emerging Treatment and Disposal Technologies" Penn State Extension Service, April 2011.

CHAPTER 9

Water Recycle and Reuse

THIS chapter discusses adding water recycle and reuse treatment systems after the existing activated sludge process. Activated sludge clarifier effluent can and has been treated for removal of TSS, TDS, chloride, silica and calcium and magnesium hardness to achieve water quality specifications for reuse in cooling towers, boilers, cleaning systems and lawn irrigation. Industrial manufacturing plants are now evaluating and implementing water conservation and reuse more than a few years ago. One of the key drivers is Corporate Sustainability Goals to reduce water use, as well as energy, greenhouse gases, emissions and carbon footprint. Water reuse is the utilization of water that has been used before for other purposes. Water recycle is the reuse of the same water one or more times for the same purpose such as recirculation cycles in cooling towers.

Figure 9.1 shows an Integrated Approach to water and wastewater treatment and reuse. The three circles showing potable water, wastewater and stormwater were typically evaluated as separate projects in the past. This has changed with the circles coming together as one integrated or holistic approach for both industrial and municipal projects. The three water sources, or circles, are evaluated together now as an Integrated Approach. As wastewater effluent permits have become more stringent to achieve better quality water with lower levels of organics, solids, metals, nutrients, etc., the treated effluent quality approaches that which can be reused in cooling towers, boilers, cleaning and wash water systems and lawn irrigation systems. Stormwater or rainwater harvesting in the lower circle evaluated for capture and reuse in cooling towers as shown in Case Study No. 4 herein. Potable water sustainability in volume and quality is another concern to industry.

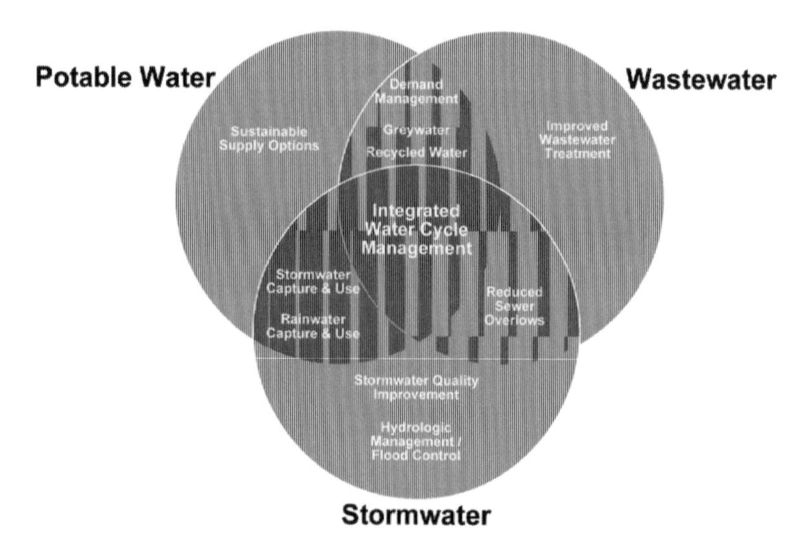

Figure 9.1. An integrated approach to water treatment and reuse.

WATER FOOTPRINT

Water footprint is the total amount of water used to make a product including the water used to make the raw ingredients and materials needed to make the product. Table 9.1 shows an example of water footprint, such as a 4 oz. cup of coffee requiring 37 gallons of water with most of this the water needed to grow the coffee beans. Similarly, a 16 oz. diet cola has a water footprint of 33 gallons with most of that to grow the sugar and other ingredients that are used to make the beverage.

The bottling plants for making cola beverages typically use 2.5 to 3 gallons of water for each gallon of product, while breweries use 4 to 5 gallons of water per gallon of beer product.

TABLE 9.1. Water Footprint Examples, Gallons of Water Used.

Product	Water Used
Apple	18 gal.
Pint of beer	20 gal.
16 oz. diet cola	33 gal.
4 oz. coffee	37 gal.
Diaper	214 gal.
1 pound chicken	467 gal.
1 pound beef	1,857 gal.
Pair of jeans	2,866 gal.
Mid-size car	39,090 gal.

TABLE 9.2. 2012 Water Stewardship Benchmarking Results, Beverage Industry Environmental Roundtable.

Product	# of Facilities Surveyed	Water Use Ratio (l/l)		Key Facts
		2009	2011	
Carbonated Soft Drink	725	2.23	2.02	
Bottled Water	131	1.55	1.47	
Brewing	296	4.53	4.00	
Distillery	80	38.55	34.55	Cooling Water Driven
Winery	27	3.78	4.74	Production Driven

Table 9.2 shows the results of a 2012 Water Stewardship Beverage Study(1) of over one thousand facilities. The water reuse per product varied from 1.5 liters per liter (l/l) for bottled water to 34.5 l/l for a distillery.

WATER BALANCE

The first step in any water recycle and reuse project is to develop a facility-wide water balance diagram. This water balance tool is essential to understand all the water inputs, both existing and potential, to the facility and how the water is used throughout the facility. It also includes how much of the water coming in ends up as wastewater discharge. Figure 9.2 shows an example of a water balance flow diagram for a pharmaceutical manufacturing facility. The key water users at the facility are shown.

Most facilities have some form of water balance to start with. Many of these water balances are outdated and need to be upgraded to existing water usages. Some of the water balances will have data gaps in which some of the flows are not available. The data gaps can be filled in by using flow monitoring measurements or estimating techniques to close the water balance.

QUALITY NEEDS

The typical end users of treated water in industrial plants are: potable water; process water; cooling towers make-up; boiler make-up; lawn irrigation; and wash water systems. Treated wastewaters are not used in the product in beverage industries nor for product or clean-in-place systems in pharmaceutical industry due to Food and Drug Administration (FDA) validation requirements. Table 9.3 shows an example of the water quality parameters typically required in cooling towers depending on the type of stainless steel. The key parameters for cooling towers and boilers are total dissolved solids, chlorides, calcium and

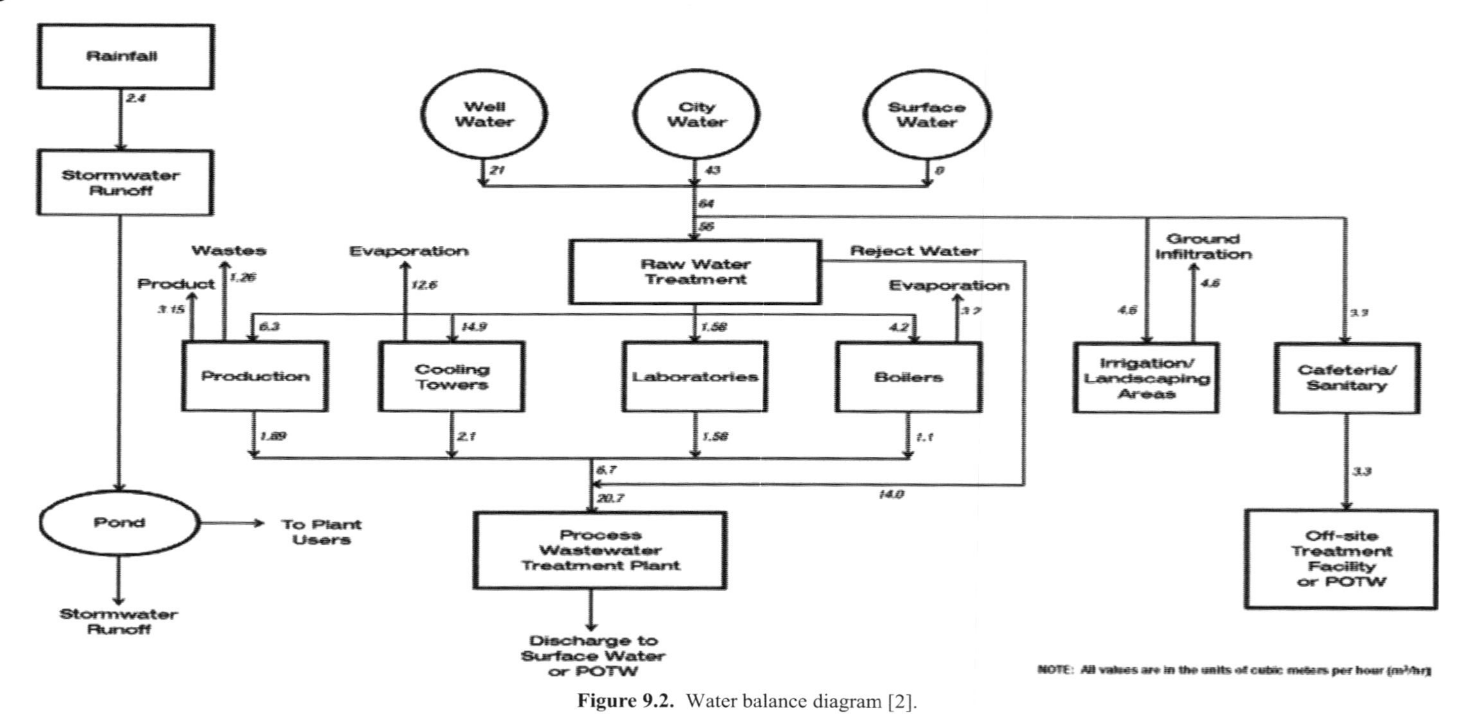

Figure 9.2. Water balance diagram [2].

TABLE 9.3. Evapco Water Chemistry Guidelines (Type 304SS Selected).

Property	G-235 Galvanized Steel	Type 304 Stainless Steel	Type 316 Stainless Steel
pH	7.0–8.8	6.0–9.5	6.0–9.5
pH During Passivation	7.0–8.0	N/A	N/A
Total Suspended Solids (ppm)*	< 25	< 25	< 25
Conductivity (Micro-mhos/cm)**	< 2,400	< 4,000	< 5,000
Alkalinity as $CaCO_3$ (ppm)	75–400	< 600	< 600
Calcium Hardness $CaCO_3$ (ppm)	50–500	< 600	< 600
Chlorides as Cl (ppm)	< 300	< 500	< 4,000
Silica (ppm)	< 150	< 150	< 150
Total Bacteria (cfu/ml)	< 10,000	< 10,000	< 10,000

*Based on standard EVAPAK™ fill.
**Based on clean metal surfaces. Accumulations of dirt, deposits or sludge will increase corrosion potential.

magnesium hardness and silica. Sampling of the treated wastewater for these parameters is typically required since they are not part of the sampling done for discharge compliance.

TREATMENT PROCESSES

Table 9.4 shows a list of the treatment processes that are used following activated sludge for water reuse and Zero Liquid Discharge (ZLD) projects. The alternatives are grouped under membrane systems for water reuse and by pressure versus electric driven processes. Technologies are also grouped by final disposal of brine. The key treatment processes used to provide the quality needed for the end user include: filtration for TSS removal, softening for removing calcium and magnesium hardness; and reverse osmosis for removal of salts such as chloride. Membrane bioreactors can be used to upgrade activated sludge by replacing the clarifier with ultrafilter membranes as a first step to achieving better quality water for reuse. The MBR reduces the TSS and associated constituents such as metals, but does not remove the TDS, chlorides or hardness. MBR followed by RO has been used for water recycle and reuse in many plants.

Case Study No. 1—Pharmaceutical Plant

The authors worked on this pharmaceutical project in 2002. The plant had an SBR system to treat about 3 MGD of pharmaceutical wastewater. The treated effluent at the time was treated further by mixed media filtration and reverse osmosis (RO). The SBR effluent was recycled back for scrubber water and the RO effluent was recycled back for use in the cooling towers. Overall, the plant was achieving about 70% reuse of treated water. The mixed media filters were eventually replaced by ultrafiltration to improve the life and replacement of the RO membranes which was less than one year due to fouling. Membrane fouling is a key concern to investigate in the design of water reuse systems. Fouling can be caused by biological slime or chemical scale build-up. Pre-treatment prior to the membranes, as well as periodic cleaning, are used to improve the life of the membranes.

Case Study No. 2—Pharmaceutical Plant

This pharmaceutical client wanted to evaluate treating the effluent from the activated sludge treatment plant for reuse in the new cooling towers. Alternatives were evaluated including: MF, UF or MBR followed by one and two pass RO. A disk and cartridge filter followed by carbon and single pass RO system was selected for design to achieve 70% water reuse. Table 9.5 shows the treated water quality required by the new cooling towers and chillers. Capital and O&M costs for this 20,000 gallon per day system were estimated at $47 per gallon, and $124 per 1,000 gallons, respectively. Figure 9.3 show the recommended disk and cartridge filters and single pan RO.

TABLE 9.4. Technologies Assessed for a ZLD Project.

Brine Minimization	Final Disposal
Pressure Driven	Mechanical
• Dual RO with Intermediate Treatment	• Vapor Compression Brine Crystallization
• High Efficiency Reverse osmosis (HERO™)	Evaporative
• Optimized Pretreatment and Unique Separation (OPUS™)	• Evaporation Ponds
• High Efficiency Electro-Pressure Membrane (HEEPM™)	Enhanced Evaporation
• Advanced Reject Recovery of Water (ARROW™)	• Wind Aided Intensified Evaporation (WAIV™)
• Seeded Slurry Precipitation and Recycle SPARRO™)	• Solar Powered Mixers
• Vibratory Shear Enhanced Process (VSEP™)	• Spray Drier
Electric Potential Driven	• Air Sparge
• Electrodialysis (ED)	Natural
• Electrodialysis Reversal (EDR)	• Salt Tolerant Wetlands (Halophytes)
• Electrodialysis Metathesis (EDM)	

TABLE 9.5. Recycle/Reuse Treatment System—Required Effluent Concentration.

Parameter	Unit	Evapco/Carrier Water Quality Guidelines	Required Makeup Water Quality (at CoC = 3) With Safety Factor = 20%
pH	su	7–9	7–9
TSS	mg/L	< 25	6.7
Conductivity	umhos/cm	< 4,000	1,067
Alkalinity as $CaCO_3$	mg/L	< 350	93
Chlorides	mg/L	< 500	133
Silica	mg/L	< 150	40
Total Bacteria	cfu/ml	< 10,000	2,667
Iron Oxides	mg/L	< 1.0	0.27
Ammonia	mg/L	< 0.5	0.13

Case Study No. 3—Food and Beverage Plant

Treatment technologies were evaluated for a food and beverage plant to achieve a high degree of water reuse including a goal of zero-liquid discharge (ZLD) for a bottling plant. Alternatives evaluated included: equalization, MF, MBR, anaerobic MBR, RO, ion exchange, evaporation and crystallization, vacuum distillation, dual media filtration and ultraviolet disinfection. Figure 9.4 shows the process flow diagram that was developed for a ZLD treatment system which included pretreatment disk and cartridge filters and carbon absorbers followed by two stage RO to achieve approximately 91% water reuse. The capital and operating and maintenance costs for this 400 gpm system were

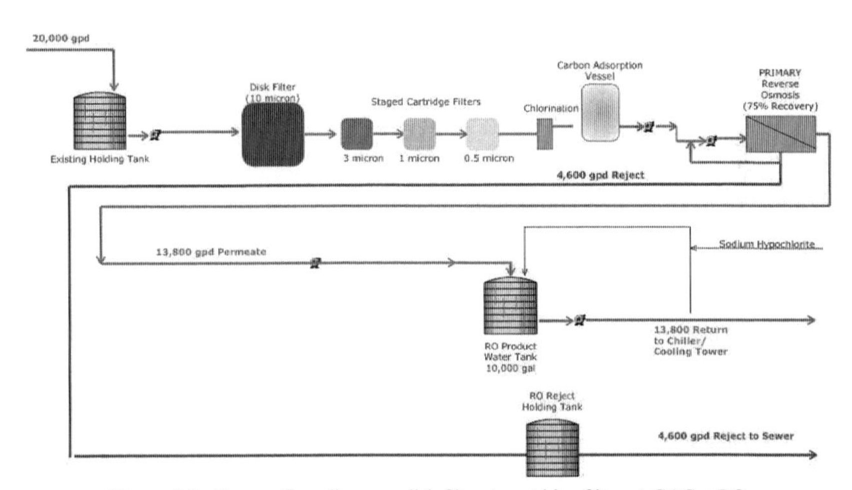

Figure 9.3. Process flow diagram: disk filter + cartridge filters + GAC + RO.

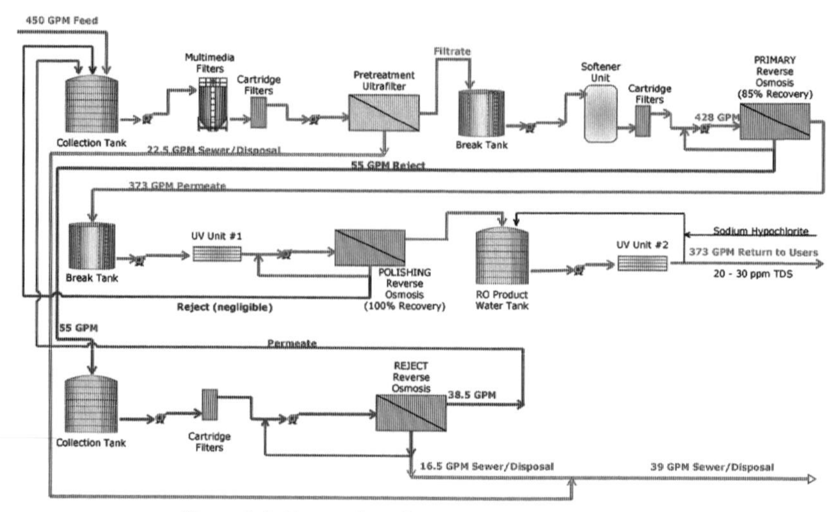

Figure 9.4. Process flow diagram: 2 UF + two stage RO.

$15 per gallon and $6 per 1,000 gallons, respectively. Key issues of concern were bio-fouling and chemical scaling fouling of the membranes. Bench-scale treatability studies were being conducted to evaluate fouling. Segregation of the initial and final bottle washing rinses was another alternative being considered with the first rinse going to an anaerobic MBR for energy recovery and the final rinse which was much cleaner to the water reuse treatment system for use as the first rinse. Table 9.6 shows the design basis flow and concentrations for the total wastewater and final rinse.

Case Study No. 3 is an example of the term Water Energy Nexus used today. Sustainability projects typically include the goal of reducing both water and energy. The challenge sometimes is that increasing the level of treatment using membrane systems increases the energy cost for running the plant with increased pumping horsepower to create the pressure needed to process water through membrane systems. This challenge can be overcome by segregation

TABLE 9.6. Design Basis—General Plant Drainage and Final Rinse from Bottle Washing.

Parameter	Unit	Bottle Washing	Final Rinse
Flow	gpm	450	250
COD	mg/L	5,000	16–393
BOD	mg/L	2,500	340
TDS	mg/L	5,000	1,832
TSS	mg/L	200	14–18
Oil & Grease	mg/L	30	1.4–3.6
pH	su	3–12	10–12

of high strength wastewaters for anaerobic treatment and biogas recovery and reuse for energy. Other energy reduction alternatives such as more efficient pumping and aeration systems can be used.

Case Study No. 4—Pharmaceutical Plant Rainwater Harvesting

This pharmaceutical company is located in Puerto Rico uses surface water (with the option to use groundwater) for the manufacture of pharmaceutical products. The company was facing uncertainty in the sustainability of the surface water supply which had multiple users and demands upstream of this plant. The local groundwater supply has quality issues due to salt water intrusion. This uncertainty presented a business risk to the production of major drugs. A site water balance was developed and water conservation and reuse alternatives were evaluated to reduce reliance on the surface water supply.

A three day site survey of the facility, interviews with plant personnel and workshops were held at the plant. A survey questionnaire was filled out prior to and during the site visit. The following tasks were included and typical of water reuse projects within the scope of this task:

- Review and refine the current and future water balance.
- Survey existing water uses at the facility.
- Identify potential water conservation and reuse projects which may have been identified before plus others identified by the project team.
- Evaluate water make up system including cooling tower and boiler.
- Evaluate potential use of surface water collected in stormwater basin.
- Evaluate potential reuse of blow downs from cooling tower and other sources.
- Identify potential reuse alternative for wastewater effluent.

The alternatives evaluation was developed based on the following:

- The current water usage at the facility was estimated to be about 134,000 gallons per day. The major water usage groups include utilities/cooling tower operations, process water usage, sanitary/cafeteria and miscellaneous usage.
- Production of a new product is anticipated to increase over the next few years from the present level of 29 percent to 95 percent by 2008.
- The impact of water reuse alternatives on wastewater treatment operations and regulatory compliance were taken into account during the evaluation process. This is especially important when considering the discharge of active pharmaceutical ingredients (APIs) from the facility.
- As part of an operating strategy/emergency management plan, the company wanted to have the capacity to continue production operations for a period of two weeks at the site in the event that either energy or water supplies are disrupted.

A site wide water balance shown in Figure 9.5, was conducted to develop an understanding of water usage and water quality at various stages and around various water usage groups. There are two PRASA water intakes at the facility. Based on available data, it was estimated that the average intake is about 123,000 gallons per day. The total water usage at the facility was projected to increase as production is ramped up. The expected maximum usage of water was about 236,000 gallons per day.

Water usage at the facility can be broadly categorized into four groups based on usage category and water quality requirements. Water usage in cooling towers/utilities represents the single largest water usage group accounting for 65 percent of the total water usage at the facility. Utilities water usage is expected to continue to remain the largest water usage group in the future (at full production capacity). Process water usage is presently at about 25 percent of the total water usage, but is likely to increase as production reaches full capacity. Sanitary usage is expected to remain constant at about 5,500 gallons per day.

Table 9.7 shows a summary of the flow and water quality parameters developed from a sampling program.

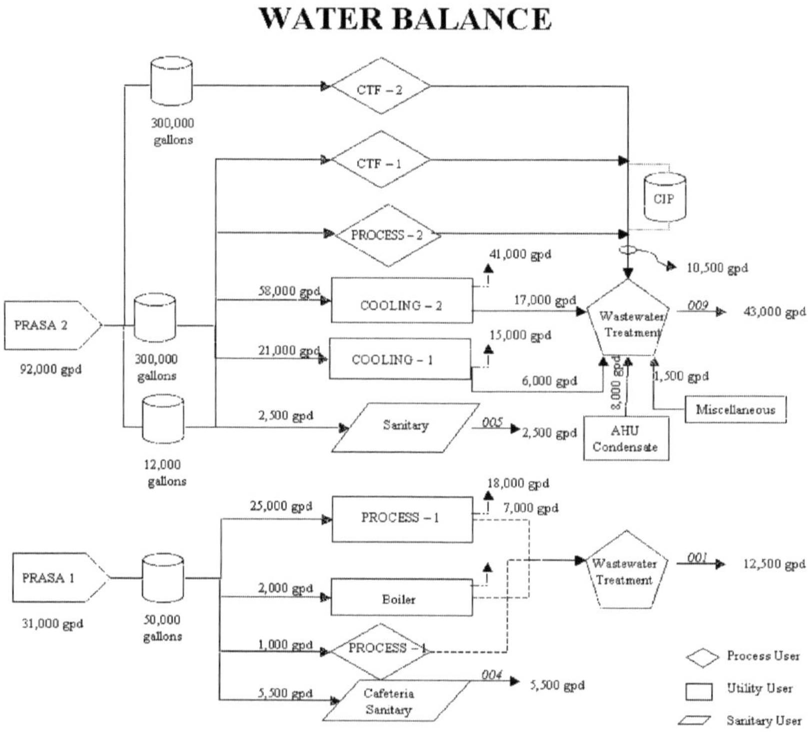

Figure 9.5. Water balance.

TABLE 9.7. Water Quality Summary.

	Type	Flow	Key Parameters			
		gallons/day	TDS mg/L	Conductivity μmhos/	Hardness mg/L	Silica mg/L
PRASA 1	Existing Source	31,000	157	274	84	24.40
PRASA 2	Existing Source	92,000	157	274	84	24.40
Well Water	Existing Source	50,000	198	317	60	21.40
Stormwater	Potential Source	55,000	176	295	68	5.57
R.O. Reject (USP)	Potential Source	10,000	388	569	4	32.80
AHU Condensate	Potential Source	8,000	9	28	4	< 0.10
CT Blow down (R.O.)	Potential Source	26,000	< 100	< 150	ND	< 0.10
Utilities/CT	Usage	134,000	157	274	84	24.40
Process	Usage	11,000	–	125	–	–
Sanitary	Usage	2,000	157	274	84	24.40
Miscellaneous	Usage	1,000	157	274	84	24.40

Approximately 134,000 gallons of make up water is required each day for the cooling towers under existing cooling tower operating conditions. At the present time the entire make-up is city water. Figure 9.6 shows the recommended alternative which achieved an 80% reduction in city water. There was potential for reuse of 8,000 gpd of AHU condensate, 10,000 gpd of reject from the USP water treatment system and 50,000 gpd of stormwater/groundwater to supplement intake of fresh water from PRASA. Water quality characterization data for the AHU condensate and stormwater/ groundwater indicated that these sources had the potential to provide high quality water with minimal pretreatment. A reverse osmosis system was proposed for the treatment of cooling tower blow down. Blending of stormwater (rainwater harvesting), groundwater, reject from the USP water treatment system and AHU condensate with city water (from PRASA) were all incorporated in the solution. The cooling tower chemistry was also optimized to achieve 6.5 cycles.

The capital cost of implementing this alternative was approximately $4 Million which included the cost for providing 14-day contingency storage for sustaining utility operations during any disruption of the water supply. The annual O&M cost for this alternative is estimated to be about $115,000 and the project payback was 3 to 4 years.

Case Study No. 5—Water Reuse in a Pork Packing Plant

A pork packaging plant wanted to expand their production, but no more water was available to them. Consequently, they approached USDA's Food Safety and Inspection Service (FSIS) about reusing their water, after tertiary

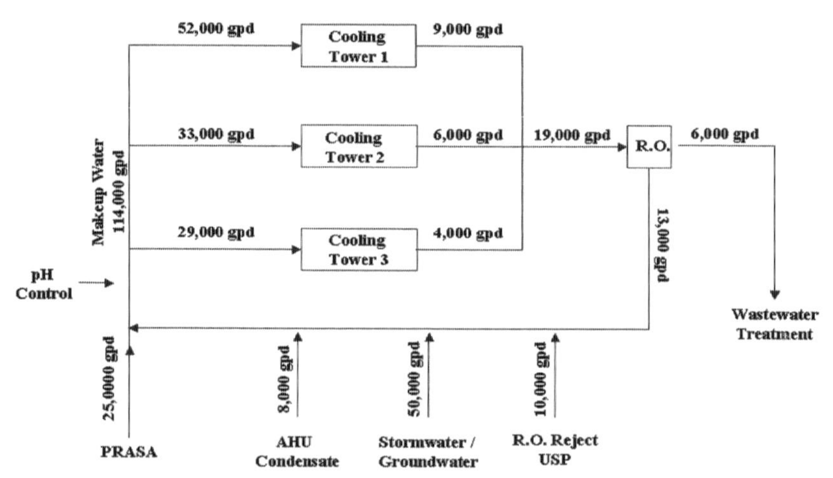

Figure 9.6. Alternative 4—existing CT operating conditions with stormwater/groundwater supplement and AHU condensate and R.O. treatment of CT blowdown (6.5 Cycles).

TABLE 9.8. Reconditioned Process Water Criteria.

Parameter	Limitation
pH	6.5–8.5 S.U.
Turbidity	< 1 NTU with no more than 5% of samples > 1 NTU and with no samples > 5 NTU
Fecal Coliform	None
Total Coliform	< 5% of samples positive
Total Plate Count	< 500 CFU/ml
Chlorine Residual	1–5 mg/l
Total Organic Carbon	< 100 mg/l

treatment, on the "dirty" side of their hog kill. A trial protocol involving extensive testing was established that would allow FSIS, as well as the company, to be certain that reuse of this water would not jeopardize the quality of the finished product. This trial dragged out and proved very costly, but after about five years, it was finally approved.

For treated wastewater to be reused, it must meet numerous criteria. The main criteria are as follows:

1. No human waste in the reuse water. This dictates segregation and separate treatment of the sanitary sewage.
2. Treatment to a level to yield a turbidity of one NTU or less. This will generally require some form of tertiary filtration, sometimes with chemical addition.
3. Maintenance of a chlorine residual throughout the distribution system.
4. Identification of reuse water piping with provisions to prevent cross connection with potable water supplies; both mechanical devices and periodic inspections.
5. Routine monitoring of the reuse water to ensure its physical, chemical and microbiological quality with provisions to terminate reconditioned water usage when problems are discovered and until they are corrected.
6. USDA approval of the protocol for use of reconditioned water.

In general, FSIS has approval responsibility for use of reconditioned water in inedible areas and the dirty side of the kill, i.e. before opening up the animal carcass. USEPA has approval responsibility for use of reconditioned water on the clean side of the kill. Hog slaughter facilities have significant opportunities for water reuse due to extensive use of water on the dirty side of the kill. As much as 30–40 percent of the total flow can be reused in some hog plants. Table 9.8 shows the reconditioned process water criteria.

Specific locations for water reuse included:

- Pens, Holding and Unloading Areas
- Inedible Rendering: barometric condensers, air scrubbers and washdown
- Wastewater Pretreatment and Treatment Facilities: washwater for screens and filters
- General: lawn sprinkling, pump seal water, cooling water for compressors and condensers
- Hog Slaughter Operations

USDA required that the reconditioned process water program also meet the following criteria:

- The reconditioned process waster must also meet EPA maximum contaminant levels for heavy metals.
- A potable water rinse on all edible product and equipment that contacts reuse water.
- A "fail-safe" system of critical control points must be in place that prevents substandard water from entering the end use part of the system that contacts product.
- The reconditioned process water lines will be physically separated from potable water lines with provisions made for potable water availability for standby use.

Figure 9.7 shows the overall process flow diagram.

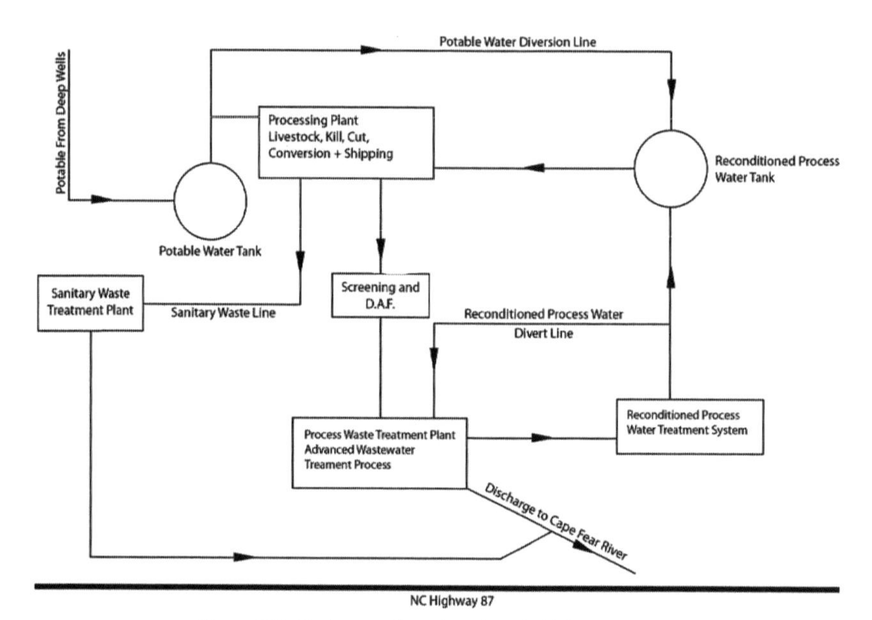

Figure 9.7. Pork manufacturing plant facilities overview.

ZERO LIQUID DISCHARGE

Several industrial plants have achieved Zero Liquid Discharge (ZLD) including a pharmaceutical plant in Puerto Rico [3] and a petrochemical plant in Mexico [4]. The economics, or payback, on most projects may preclude the selection of ZLD system but it continues to be evaluated as an alternative on water recycle and reuse projects to achieve sustainability, environmental compliance and business risk reduction goals. ZLD alternatives include ultrafiltration and reverse osmosis (UF/RO) to remove total dissolved solids and chlorine for reuse both in cooling towers and boilers. The RO reject or brine can be sent to a brine concentrator or to evaporation and crystallation treatment systems to reduce the water discharge further and approach ZLD.

Beverage Industry

Water reuse projects are being implemented in the beverage industry for bottling plants. Wastewater from bottle washing is almost 50% of the total wastewater in a bottling plant and the final bottle rinse can be reused in the bottle initial pre-rinse step [5]. Average characteristics of the final bottle wash rinse were as follows: Turbidity 40 NTU, COD 48 mg/L, TSS 56 mg/L and TS 694 mg/L. The results of treatability tests showed that the final rinse water can be used in the pre-rinse and pre-washing after removing all of the suspended solids, 80% of the BOD and 75% of the dissolved solids. This can be accomplished using filtration, carbon adsorption, reverse osmosis, or filtration, adsorption and ion exchange. The installation of these treatment technologies in the soft drink industry would decrease bottle washing water consumption by 50%.

The overall water balance for this bottling plant revealed that 76% of the raw water consumed daily ends up in the biological wastewater treatment plant [6] and the ratio of raw water to product was approximately 4.1 to 1. A large portion (40%) of this wastewater is generated from the bottle-washing units. By employing microfiltration for polishing of the WWTP effluent, the plant can recover process water for reuse such that, groundwater input is reduced by 40% and liquid discharged to the receiving water is reduced by 56%. A microfiltration-reverse osmosis system was proposed to purify the rinse water for reuse in the bottle washing process, thereby reducing raw water consumption further to 58% and the liquid discharge by 81.5%. A dual filter media-ion exchange system was evaluated to reduce raw water input to 57% and the liquid discharge by 80.5%. The dual filter media ion exchange capital costs were approximately 50% of the MF + RO system.

Another bottling plant pre-treats the raw water on-site to meet product quality requirements before being used in the manufacturing process. The water treatment included softened water for bottle washing [7].The bottle wash uses approximately 62% of water daily. The purifying process of a bottle-cleaning consists of several baths and rinse stations, followed by the main cleaning in a lye-bath which contains 2.0–2.5% NaOH at a temperature of about 80°C. After leaving the lye-zone the bottles are cleaned and disinfected. There are several rinse stations that remove the lye from the bottles and cool them step by step. After leaving the cool water bath at 28°C, the water goes to the pre-cleaning zone and from there to the sewer. A recycling system in the rinse stations was evaluated to reduce the amount of fresh water used. The volume of wastewater generated by the processes varies between 63,000 to 150,000 gpd. The effluent represents approximately 56% to 71% of the total water used. The bottle washing generates approximately 62% of the wastewater and operates 16 hours a day. The second activity which contributes to the wastewater is the CIP cleaning system. The clean and rinse of syrup tank preparation and the bottling machines use hot water with 2.5% caustic and hot water at 80°C, followed by cool water. This operation takes place at the beginning and end of each production, several times a day.

This paper [8] presented an overview of the use of MBR technology for industrial wastewater applications including the selection of the optimal MBR system. The paper also briefly discussed two (2) full scale case studies using MBR systems; one at a Nestle facility in New Milford, Connecticut and the second at a General Mills plant in Covington, Georgia. The General Mills plant included RO after the MBR system for water reuse. A comparison of the conventional external MBR configuration and the internal MBR design is presented, including a summary comparison table and a table of vendors for both MBR systems. A discussion of the effectiveness of an internal MBR installed to treat starchy waste at the Nestle facility in New Milford, CT is also presented.

This soft drink company had implemented a biochemical oxygen demand (BOD) load reduction program by segregating at least 70% of the BOD wastewater streams and then treating the segregated wastewater to at least 90% BOD removal [9]. Various wastewater treatment technologies were evaluated to treat this high strength organic load. A bench scale pilot study was conducted to evaluate the MBR process relative to its effectiveness and to develop process design parameters for the treatment system. The paper described the laboratory pilot study designed to test the feasibility of using a thermophilic MBR to treat soft drink bottling waste. The process wastewater came from a can crushing operations at one of the bottling plants. It was concluded that a thermophilic MBR would successfully treat soft drink bottling waste (COD removal = 96% and $CBOD_5$ = 99%).

This paper discusses the treatment steps required for a wastewater discharge for a large bottling plant in Puerto Rico [10]. The bottling facility had previously hauled its strong wastewater (14,000 mg/L COD) at a wastewater flow of 70,000 gpd off-site. The local treatment plant had the capacity for the flow from the bottling plant, but not the organic loading. The bottling plant decided to investigate pretreatment options for a discharge into the publicly owned treatment works including the design, construction, start-up, and operation of the MBR pretreatment system. A MBR system was selected over diffused aeration & clarification and SBR. The treatment train consists of an Equalization Tank, followed by Urea and Phosphorus feeds, and then two (2) MBRs. The treated effluent from the MBR was initially used for water reuse operations for the boilers, cooling towers, compressors, bottle warmers, lubrication of production line conveyors, and cleaning operations for wastewater system. High total dissolved solids (TDS) in the MBR effluent required the implementation of a RO system following the MBR system to remove the TDS. Startup difficulties pertaining to nutrient balance causing biofilm production and membrane fouling is also discussed.

This paper describes the use of MBRs for treating industrial wastewaters [11]. Three (3) examples of the use of MBR technology were presented. The full scale case studies included a corn wet milling facility in the Midwest; a Nestle's beverage plant in Anderson, Indiana, and a Smucker's Crisco plant in Cincinnati, Ohio. The paper summarized the design basis for the three (3) plants including expected effluent characteristics from the MBR systems. The paper indicated that one of the advantages of the MBR system is that the treated effluent can be used for water reuse. However, there was no discussion of operation areas that could be used for water reuse that additional technologies (e.g., RO) may be required for selected areas for water reuse or the expected percent capture of the treated effluent for reuse.

Frito-Lay's manufacturing plant in Casa Grande, Arizona, makes snacks including corn and potato products (Lay's, Ruffles, Doritos, Tostitos, Fritos and SunChips). In the arid region of the southwest U.S., Frito-Lay completed a project with the ambitious goal to run the plant almost entirely on renewable energy and reclaimed water while producing nearly zero waste—something the company refers to as "Near Net Zero" [12] The average daily design flow of the PWRTP is 0.648 mgd (28 L/s) from the production facility; characteristics of the influent are biochemical oxygen demand of 2,006 mg/L and total dissolved solids of 2,468 mg/L. All sanitary wastes (i.e., bathroom connections) are segregated and discharged to the city sanitary sewer for conventional treatment at the City of Casa Grande Wastewater Treatment Plant.

The reuse quality established by Frito-Lay/PepsiCo required the water to meet EPA primary and secondary drinking water standards. The process water that is used to move and wash potatoes and corn, clean production equipment, and for other inplant cleaning and production needs, is reclaimed for reuse in the process. The reclaimed water quality from the PWRTP is of higher quality than the local potable water supply in terms of alkalinity, arsenic, and silica. The treatment process includes: internal-feed rotary drum screening, equalization with pH adjustment using carbon dioxide, primary clarification/sedimentation, activated sludge with biological nutrient (nitrogen) removal in MBR,

TABLE 9.9. Current Brine Disposal Methods [12].

Disposal Method	Limitations	Percent of Systems Currently Practicing	Controlling Regulation
Surface Water Discharge	• Sufficient assimilative capacity must exist in water body to accept discharge	47%	Clean Water Act
Municipal Sewer (POTW)	• POTW has sufficient capacity to accept flow • Must meet POTW discharge permit	24%	Clean Water Act
Deep Well Injection	• Existing drinking water aquifers must be protected • Suitable local geology must exist	17%	Safe Drinking Water Act
Land Application	• Existing drinking water aquifers must be protected • Suitable soil characteristics	7%	Safe Drinking Water Act
Evaporation Pond	• Large tracts of land available • High evaporation rates • Existing drinking water aquifers must be protected	4%	Safe Drinking Water Act
Recycle for Industrial Use	• Highly specific to project	1%	Depends on Use

granular activated carbon (GAC), UV disinfection, reverse osmosis, and chlorine disinfection prior to reuse. Treated water is stored in a 200,000 gallon storage tank. The GAC system was added in 2011 to enhance treatment for additional recovery and to further protect the membranes. Reject water from the LPRO is discharged to the city of Casa Grande Wastewater Treatment Plant.

Brine Disposal

Disposal of the brine solution or RO reject water is a key issue in the ZLD alternatives evaluation and system design. Table 9.9 presents a summary of the brine disposal alternative being used [13]. The disposal method most used are discharge to surface water and discharge to the municipal sewer or POTW. On ZLD projects, the brine is typically treated using evaporation and crystallization processes to eliminate the water discharge and create a salt waste requiring disposal. Some landfill leachate projects have recently selected RO treatment with brine disposal back to the landfill.

In summary, most wastewater treatment plant upgrade projects now include looking at future water recycle and reuse in the alternative evaluations. Sustainability metrics and triple bottom line benefits (social, economic and environmental) and performance are being driven for industry to approve projects. The technologies are certainly available to treat effluent from wastewater plants for recycle and reuse for various water uses in a facility.

REFERENCES

1. Christenson, T. 2012. "Water Use Benchmarking in the Beverage Industry, Trends and Observations."
2. GEMI® Water Sustainability Planner, 2007.
3. Plant visit in Puerto Rico, 2000.
4. Wong, Joseph M. 2011. "Membranes for Wastewater Reclamation and Reuse for Petrochemical and Petroleum Refining Industries." WEFTEC 2011, Los Angeles, CA.
5. Camperos, Ramirez E., P. Migaylova Nacheva and E. Diaz Tapia. "Treatment Techniques for the Recycling of Bottle Washing Water in the Soft Drinks Industry." Mexican Institute of Water Technology, Paseo Cuaujmajiac 8532, Progreso, Juitepec, Morelos, 62550, Mexico.
6. Visvanathan, C. and Anna Marie M. Hufernia. "Exploring Zero Discharge Potentials for the Sustainabilitty of a Bottle Washing Plant." Environmental Engineering Program, Asian Institute of Technology, P.O. Box 4, Klongluang Pathumthani 12120, Thailand.
7. Hsine1, E. Ait, A. Benhammou1 and M-N. Pons2. "Industrial Water Demand Management and Cleaner Production Potential: A Case of Beverage Industry in Marrekech—Morocco., 1Laboratory of Automatics and Processing." University Caddi Ayyad, Faculty of Science Semlalia, Avenue Prince Moulay Abdellah, P.O. Box 2390, Marrakech—Morocco, 2Laboratory of Chemical Engineering Science, CBRS-ENSIC-INPL 1, rue Grandville, BP 4451, F-54001 Nancy cedex—France.
8. Sutton, P.M. 2006. "Membrane Bioreactors for Industrial Wastewater Treatment: Applicability and Selection of Optimal System Configuration." *Proceeding of the Water Environment Federations 77th Annual Conference,* October 2006.

9. Togna, A.P., Y. Yang, P.M. Sutton and H.D. Voigt. 2003. "Testing and Process Design of a Thermophilic Membrane Biological Reactor to Treat High-Strength Beverage Wastewater." *Proceedings of the Water Environment Federation 74th Annual Conference,* October 2003.
10. Murray, C.W., Abreu, L.H., Husband, J.H. 2005. "Treatment of a High Concentration Beverage Waste Stream Using MBR Technology." *Proceedings of the Water Environment Federation 76th Annual Conference,* October 2005.
11. Peeters, J., Sparkes, A., Daumgarten, S. 2008. "Solving Industrial Water Resource Management Challenges with MBR Technology." *Proceedings of the Water Environment Federation 79th Annual Conference,* October 2008.
12. Haghighipodeh, Mohammed R. and Al Goodman. "Evaluation and Lessons Learned from Full Scale Water Recovery and Reuse Project at a Food Manufacturing Plant. *WEFTEC 2012,* New Orleans, LA.
13. Mickley, M. 2012. "U.S. Municipal Desalination Plant Statistics and Concentrate Management Practices." 2012 Membrane Technology Conference and Exposition. Glendale, AZ.

Index